By the same author

THE ADVENTUROUS GARDENER
FOLIAGE PLANTS
THE MIXED BORDER
THE WELL-CHOSEN GARDEN
THE WELL-TEMPERED GARDEN
A YEAR AT GREAT DIXTER

CHRISTOPHER LLOYD

Clematis

REVISED WITH
TOM BENNETT

**Capability's
Books**
Deer Park, WI
54007

First published in Great Britain by *Country Life* 1965
Revised edition published by Collins 1977, 1979
This edition, extensively revised, first published in the USA in 1989 by
Capability's Books
P.O. Box 114
Deer Park, WI 54007

Copyright © Christopher Lloyd, 1965, 1977, 1979
Copyright © Christopher Lloyd and Tom Bennett, 1989

All rights reserved

Printed and bound in Great Britain

ISBN: 0-913643-04-1

Library of Congress Catalog Card Number: 88-63334

Contents

List of Illustrations

All photographs were taken by
Christopher Lloyd and Tom Bennett

Forewords
to 1989 edition

Large sections of any book on a popular florist's flower are bound to become out of date or inadequate quite quickly. Conscientious revision is a laborious and often tedious job, and I have revised *Clematis* so often that I felt a weight upon me at the thought of doing it yet again. Writing a new book on a new subject is so much more fun. Yet the demand was there. Could I possibly find someone to take on the main burden of the job for me?

Tom Bennett came importantly into my life on 9 September 1987 (although, on looking up my records, I found that my nursery had posted him an order for two clematis on 30 September 1968, when he was fifteen years old, so his interest, like mine, started early). He visited my gardens and brought me a *Clematis* 'Lambton Park' to try. I was on my knees weeding and continued doing so while we conversed. After a few minutes, I asked if he would consider revising my book. We went on from there and Tom has become one of my closest friends.

He has done all the donkey work but has always been in close liaison with me, and has ever remained considerate of this still being my book. We have gone through every alteration and addition together. One day he will write his own clematis book and I should count it a privilege if I were allowed to look over his shoulder at the manuscript and make the odd suggestion; for instance, of where one word might be allowed to do the work of two.

That seems a good point at which to hand over!

C.L.

September 1988

I count myself fortunate to have been involved in this revision. My mother bought me the first edition of *Clematis* twenty years ago, when I was fifteen, and I can still recall my delight as I discovered that clear, informative and frequently witty style for which Christopher Lloyd has become famous. It is a timeless book, able to help no matter how the subject is approached, and my well-thumbed, though now dog-eared, copy is never dusty. In 1977 he achieved the near impossible by writing a revision which was as good as, yet entirely different from, the original. It was a pity that it quickly went out of print.

I first met Christopher in September 1987 during a visit to Great Dixter, his home and garden in Sussex, and I asked him when his book might be revised and reissued. It was long overdue; a great many new hybrids and unfamiliar species had entered the lists and, in any case, his book was sorely missed. After the briefest of pauses and quite out of the blue, he suggested that I might do it. I was taken aback at such responsibility but it was not one that I had to shoulder alone. In the event, many people contributed and I thank them all. I owe them much; their help and support were invaluable.

Jim Fisk claims to have retired but. is greatly exaggerating. He answered all my queries promptly, and sent slides of new crosses made in Poland and some plants that I might see them at first hand. Alister Keay in New Zealand gave details of his introductions and useful information about the species found there. I was able to see some of these for myself when Dr Jack Elliott kindly opened up his Kentish garden to Christopher and me one particularly raw March day. Richard Pennell, the late Walter Pennell's son, kept me abreast of what was happening at his nursery in Lincoln. Roy Lancaster put me on the right lines towards finding more about *C. florida*, and Dr Alan Leslie, Registration Officer with the Royal Horticultural Society at Wisley, delved even deeper, both into that species and others. I owe Alan a particular debt; he also checked the chapter on hybridization and gave encouragement when it was most needed. He is a good friend.

I made another, very special, friend during the course of this revision. John Treasure and Christopher Lloyd have been close for over thirty-five years, and each knows as much as the other when it comes to clematis. John asked me to stay at his home, Burford House, in Shropshire, so that I might look through notes and documents he has and talk to the staff who now run the nursery. They were all kind and generous in their help. John was intensely interested in what I was doing and I have made a number of visits to Burford since. Each has been a delight. I thank him for his help, his friendship and the indulgent smile he gave when my three small children performed energetic cart-wheels over his close-cut lawns the last time I visited.

For any author to trust his work to another must be a traumatic experience. Throughout the course of my work on his book, Christopher gave no real hint of anxiety or impatience even though I knew that, had the positions been reversed, I would have been very tetchy indeed. My frequent stays at Great Dixter, ostensibly to review progress, were anything but difficult. He is a superlative host (and an excellent cook) and I was made to feel totally at ease. I have retained much of the original book simply because it could not be bettered; any alterations and additions that needed to be done, we did together. That it happened this way was due to Christopher's kindness, patience and friendship and I could not have asked for a more pleasurable task.

Valerie, my wife and best prop, has put up with my frequent absences and all the paraphernalia involved in writing – reference books, cata-logues, letters, discarded drafts, all littering the house. The dining table was lost to it for nine months. She gave me endless cups of coffee, lots of encouragement, and hardly a cross word.

T.H.B.

September 1988

Preface

Having finished his book, the author writes his preface. There seem to be an unusually large number of people to mention in this one, making it difficult to know whom to start with, so I'll start with myself. I wrote a book, *Clematis*, for *Country Life*, which they published in 1965. When they were taken over by the great IPC empire, soon to become part of the still larger Reed's group, all such monographs serving small minority interests were chopped and *Clematis* went out of print. Collins, who had handled *The Well-Tempered Garden* and *Foliage Plants*, were enthusiastic about reissuing *Clematis* in whatever form I chose, and at first I favoured dishing up the old text with a few cuts here and a few minor additions or alterations there.

> Thrift, thrift, Horatio! the funeral bak'd meats
> Did coldly furnish forth the marriage tables.

The more I thought about it, the less it seemed to me that this would do. No one who is interested in his subject can remain in the same spot. Ideas, opinions and values are changing all the time. Here he discovers that he was wrong, there that a whole vein of new material is waiting to be quarried. On the technical side, advances are being made altering our entire approach to (most notably, in this case) propagation and to the control of diseases. In the end, although I used the same framework of chapter subjects, I re-wrote the entire book except for Chapter 3. I intended to re-write that and started to do so. It went stickily. The black feeling of having said it all before brooded heavily on me. I turned up the old chapter and made the disconcerting discovery that it was vastly superior to my present efforts. Chapter 6 was the one I most enjoyed. It is the longest and gave me the most trouble in its assembling, but the ferret leads an exciting life even though there cannot be a rabbit at the end of every passage. And the next two chapters, on planting and pruning, seemed to go well, written at high speed and with daunting illegibility.

Having devoted so much thought to this one genus, I now feel that it has grabbed a disproportionate share of my gardening attention. There are so many other good plants; why specialize in one? Yet the book needed to be written, as it will, in the future, need to be frequently revised. What other clematis literature has the fanatic to turn to? In 1979, Raymond Evison's slim volume, *Making the Most of Clematis*, showed many enticing pictures but was light on in-depth information, particularly regarding propagation and disease. In 1975 Jim Fisk brought out *his* second volume on the subject, calling it *The Queen of Climbers*. Published privately, it was available from his nursery at Westleton in Suffolk but is now out of print. It contained many splendid colour photographs and much good sense. Also a lot about Jim Fisk and his business. It is always delightful to write about oneself. I never tire of it. But his book was short; it did not attempt or pretend to cover the subject in detail. His Wisley handbook, lately revised, is tied more closely to the subject and is nicely illustrated. The only other handbook around at the moment is *Clematis* by Ethne Reuss Clarke, published by Aura Editions. It is lavishly illustrated although, as often happens, some of the colours are very poor. The text too may mislead. 'Jackmanii Alba' is described as 'not a particularly vigorous hybrid'. Anyone relying on that information would be in difficulties in no time. The International Clematis Society, formed in 1984, publishes a *Journal* and is a new source of information although, latterly, its publication has been erratic.

Stanley Whitehead's *Garden Clematis* came out in 1959 and was reprinted; it may still be in print. It always gave me the impression that the publisher had said, 'We need a book on clematis in our Gardening series, Dr Whitehead; would that be up your street?' And he had replied, 'Clematis? Ah yes! Clematis. Well, I think I might be able to tackle that', and went on from there.

I remember how thrilled I was when Ernest Markham (William Robinson's head gardener at Gravetye) brought out his little book *Clematis* (also published by *Country Life*) in 1935. That reached a second edition in 1939, after his death, and was revised by A. G. L. Hellyer in 1951. But it was only a little book. The subject of propagation from cuttings, for instance, was discussed in just over one page. His descriptions of cultivars were spartan: 'Lasurstern – deep blue; very large flower.' 'Beauty of Worcester – violet-blue.' Difficult to be kindled by that.

Markham's employer, William Robinson, wrote an even slimmer book, *The Virgin's Bower*, which was published in 1912 and ran to only 35 pages (Markham managed 110). Its full title was *Clematis – Climbing Kinds and Their Culture at Gravetye Manor*, and its chief interest lies in giving an insight into attitudes prevailing at Gravetye before they acquired Morel's hybrids – an event of some significance which I shall return to several times later in this book. Robinson irritates by using the contrived epithet 'Virgin's Bower' for every species he describes, a habit that was taken up by Markham and is perpetuated by some of today's writers and growers. Robinson detests the doubles; on p. 26 he insists: 'I never plant one if I know it. The wondrous grace of buds and flowers are lost by doubling ... Sometimes if a double kind comes in by mistake, I get it on the fire soon.' Clearly, many of his contemporaries disagreed, *vide* 'Beauty of Worcester', 'Proteus', 'Duchess of Edinburgh', 'Countess of Lovelace' and many more. These would hardly have survived had his views been dominant. Despite many ill-considered views (on grafting and the causes of disease included) he showed great imagination in the ways that he suggested growing clematis, now widely taken for granted, such as through trees and shrubs and in juxtaposition with other garden plants.

Before Markham (Robinson notwithstanding) we had, for sustained reading, to go right back to the first book ever written on our chosen flower, *The Clematis* by Thomas Moore and George Jackman, 1st edition 1872, 2nd edition 1877. A. G. Jackman made extensive notes in his copy of the latter for a revised edition some time early this century. It never came to fruition, but Rowland Jackman (who, alas, is no longer with us; he died soon after completion of my script) was most generous in lending me this copy for as long as I needed it, and as I have taken three winters over this book that must be for more than two years. It has made an invaluable link of the past with the present and I have made constant reference to Moore and Jackman in the text. Also to Bean, which means *Trees and Shrubs Hardy in the British Isles* by W. J. Bean, who was Director of the Royal Botanic Gardens, Kew. I have used both his original 1914 edition (which my parents owned) and the 8th and latest revision of 1970 under Sir George Taylor's general editorship, Mr Desmond Clarke having shouldered the lion's share of the burden. They have kept me more or less on the straight and narrow in so far as they were able. In return I will point out to them that the flowers of *Clematis montana* are not solitary but in axillary clusters; and that *C.* × *durandii* and *C.* × *jouiniana* are not climbers within the generally accepted meaning of the word.

My burrowings kept me for a good many hours in the RHS Lindley library in Vincent Square, but it was the late Mr Hugh Thompson who put me on to so many meaty references there; he was a very great help in many ways, as the reader will be made aware hereafter. He used to write in his small neat hand with blue ink on deep blue paper (as John Treasure does; there might be an association worth pursuing here, with a love of deep blue clematis), or else he'd cram into a postcard as much as any normal letter-writer would spread over two pages. Alas, that leukaemia should have taken him untimely, before my book was out. I've quoted from him freely because the way he expressed himself was always so good-humouredly humorous.

Everyone to whom I have turned for help has been extraordinarily generous and co-operative. A visit to the House of Pennell in Lincoln informed me not only of the latest crosses coming to fruition but also that Walter Pennell was a high-powered amateur astronomer, with an impressive telescope occupying that key position in his garden where lesser mortals would site the bird bath. Finally I begged to be shown more and yet more of his astral photography in preference to the nurseryman's equivalent of 'here's another of mum and the kids; she's paddling in that one' – I mean the coloured transparencies of his own clematis crosses. It is sad to record that he died in the summer of 1977, soon after he had sent me an appreciative letter regarding this book's first edition.

I wasn't able to visit Mr Fisk but he did his best to answer my queries.

Most of all I owe to the House of Treasure. John Treasure has been a close friend since 1952, and we learned about clematis together or separately by interchange of ideas and information, step by step right from the outset.

Raymond Evison, whom I have known since he was a boy, has helped me unstintingly, especially on his methods of propagation and also on the endlessly teasing questions of variety identification and naming, what different nurseries are calling the same plant, what new species and cultivars have reached him, and all such details. He is a highly professional and efficient person so I am indeed lucky, here.

Chapter 1
THE APPEAL OF CLEMATIS

E very gardener loves clematis. I have yet to meet the man or
woman who was not attracted by them. There may often be
certain categories of which they disapprove – the stripy cartwheels of
'Nelly Moser', for instance – but these will be outweighed by the rest.
In few florist's flowers can such popularity be taken for granted.
Think of all the people who loathe chrysanthemums, dahlias, gladioli,
rhododendrons, cacti and calceolarias. Mistakenly, of course, but there
it is. Yet clematis are universally admired.

I must discuss why this should be at the start, because unless I can
convince myself as well as you that this flower's popularity is based on
sterling merits, there can be little excuse for writing a book on it.

Size of bloom will be the first aspect to touch us. Sheer size to make
us gape. And colour: sumptuous velvety lavender, crimson and purple.
And then texture. If we want to touch a flower as well as feast our eyes
on it, that is a sure sign that it has us in its emotional grip. We want
to enjoy it through our finger-tips. That such size should be linked
with such delicacy, such fragility, is altogether irresistible.

I should like, at this point, to be able to invite the reader to bury his
nose in a bloom and inhale, thus bringing yet another of our senses
into satisfied play. But this sense must be kept waiting: no large-
flowered clematis can assume the role of a full and deep red rose (many
of which, be it noted in an aside, are sadly lacking in the scent that
should be their essence). But if we look again we can revel in its
shapeliness, in the generous overlap at their base of sepals that undulate
and gradually taper until they expire in fine points; in the central boss
of stamens wherein pale filaments are often contrasted with rich red-
purple anthers. And now, since we approached our bloom with over-
hasty enthusiasm (trampling on half a dozen lowly and disregarded
victims in our eager progress), we should stand back a little (trampling
on another half dozen lowly disregarded plants in our single-minded
absorption) in order to admire the habit of our clematis, the way in

which it has threaded its passage over a supporting shrub (assuming for the moment that it has been given the benefit of this most flattering as well as natural type of support) and has distributed its blooms in swags, ropes and garlands, as though intentionally decorating and celebrating a festive occasion.

So far we have been considering a large-flowered clematis, for it is the very size of its blooms that will make its first impact on a child (it certainly did me from the age of ten or earlier) or on the similarly unsophisticated adult on whom flowers and gardening have broken like a revelation, as often happens, for instance, when a newly married couple find themselves in possession of their own garden for the first time.

But it is not long before we come under the sway of the small-flowered clematis that make their mark by the quantities of blossom they carry instead of by their individual size. And this is where our sense of smell is brought gratifyingly into action. Vanilla, hot chocolate, almonds, lemons, primroses, cowslips, violets: all these and other essences are wafted towards us on a light and teasing current of air so that we stop in delighted perplexity and seek out the cause with exploratory eyes, looking up and down and round about us. Then light upon a clematis such as montana or flammula, and follow the first spied cluster of flowers quickly expanding into a great pool, a foaming sea of blossom, spraying in every direction and finally flinging itself into space. Small wonder we are exhilarated.

Such, then, is the burden of my song. We shall examine and revel in many more fascinating details as we pursue our subject, and we shall also become painfully aware of the flies in the ointment. But I hope that at the end there will be no doubt remaining that clematis are a fascinating and varied group of flowers, and that in taking it to their heart, a large public have shown their natural good taste in grateful acknowledgement.

Chapter 2
BACKGROUND TO CLEMATIS

I shall not give a technical definition of clematis; that can easily be looked up in any flora or gardening dictionary. But there are a number of botanical features with regard to the genus that are of interest.

It belongs to the family *Ranunculaceae* (of the buttercups, *Ranunculus*) and this includes a rich selection of good garden plants, some of which cannot help but remind us, in their flowers or foliage, of clematis. A panicle of *Thalictrum delavayi*, for instance, could almost be a small-flowered clematis, while single and double anemones are strongly suggestive of the larger-flowered types. A further similarity between this and some other members of the family such as *Caltha* (king-cup), *Thalictrum, Anemone, Delphinium* and *Aconitum* is that the calyx (made up of sepals) comprises the showy, colourful part of the flower while the corolla (made up of petals) is either inconspicuous or absent. In *Clematis* it is absent with the possible exception of the atragene group – *C. alpina* and *C. macropetala* – in which the 'doubling' of the flower may be ascribed either to petals or to petal-like stamens or to both.

The fruits of a clematis are worthy of note, since the styles which surmount them have continued to grow after the flower's fading and, with their plumed tails, often contribute a pleasing effect as the seeds ripen. In our native *C. vitalba* the styles become feathery and pale grey, and are aptly known as Old Man's Beard.

Clematis are exceptional, in their family, for being mainly (though not always) woody. And most clematis (though not all) are climbers. Their leaf stalks (petioles) are adapted for this purpose to twist round any suitable aids to support with which they come into contact. But they cannot get a grip on a flat surface, like a wall, as can the ivies and virginia creepers.

There are said to be some 300 species of clematis, of cosmopolitan distribution but concentrated primarily in the cool-temperate northern hemisphere. A large number are supremely uninteresting from a horticultural viewpoint. Most clematis are deciduous, but some of the less hardy species from warmer climates are evergreen. There is no need for an ugly rush to acquire these, as the British climate sadly batters their foliage between autumn and spring. Ten of them hail from New Zealand, and an interesting point in respect of this little nucleus is that they are all dioecious (male and female organs on separate plants), whereas most clematis are hermaphrodite.

There has at times been a move among certain systematic botanists to split two sections of the genus *Clematis* off into separate genera: *Atragene* and *Viorna*. The former includes *C. alpina,* its Chinese equivalent *C. macropetala* and its North American ditto, *C. verticillaris* (syn. *C. occidentalis* var. *occidentalis*). It has petal-like processes (petaloid staminodes) located between the flower's real stamens and its sepals. *C. viorna* is the type of *Viornae,* characterized by a leathery, urn-shaped flower in which the four sepals are joined (connate) at the base, opening out separately at their tips. Most, if not all, of this group are North American and include the important red-flowered *C. texensis.* The European *C. integrifolia* is sometimes also included.

Atragene and *Viorna* are still included with *Clematis* in most works of reference, and are so here.

Clematis have been grown in English gardens since at least the end of the sixteenth century. Double forms of *C. viticella* were already mentioned by Gerard and Parkinson. Even so, little interest was taken in them until they began to be crossed and 'improved' in the 1850s. Thereafter, for the next twenty years, an extraordinary spate of hybridizing activity was let loose throughout Western Europe. This was not peculiar to clematis alone but to all florist's flowers, and was stimulated by the flood of new introductions of species and hybrids arriving at that time from all over the world. A large proportion of the clematis we grow today were created in those years.

Nowadays it is rare for a clematis to be put up for award at the Royal Horticultural Society's shows, but in the 1860s, 70s and 80s it was happening all the time and the award given was the First Class Certificate (FCC). The lower-graded Award of Merit (AM) was not instituted until 1888, so it is fair to say that the awarding committees exact a higher standard for the FCC today than they did then (though it should be added that the whole system of RHS awards is in process of being overhauled at the time of going to press). On the other hand the standard of cultivation was probably higher in those days than now. More care was lavished on the individual plant. Labour was cheap.

The new, large-flowered clematis derived from three Far Eastern species – *C. florida, C. patens, C. lanuginosa* – and one European, *C. viticella* (with another, the herbaceous *C. integrifolia*, peeping over its shoulder).

C. florida is a native of Hupeh (W. China), but no one living has ever reported seeing it there, or anywhere else for that matter. Roy Lancaster, a traveller in those parts, wrote Tom that he had never seen it or anything under this name. He added that in the past *C. courtoisii* had been referred to as *C. florida,* and *C. cadmia* is related. Some information has come to light, however: it is described in *Flora Reipublica Popularis Sinicae* Vol. 28 (1980), and illustrated with a detailed line drawing. This shows that it differs from *C. patens*, which might confound those who consider the two species synonymous. China has opened up considerably to botanists and others in the last decade, so maybe, before long, *C. florida* will reappear from the wilderness.

In its double-flowered cultivars it has long been grown in Japan, and it was from there introduced to Europe by Thunberg in 1776. This was most probably the double greenish-white cultivar 'Alba Plena'. Such, at any rate, was the opinion given in Volume 22 of *Curtis's Botanical Magazine,* published in 1805.

The other, even better known, cultivar of *C. florida* is 'Sieboldiana', also known variously as *C. sieboldii* and *C. florida bicolor*. This was introduced from Japan by Siebold and came to England from his nursery in 1836.

The status of *C. patens* has changed recently. In the 1984 International Clematis Society *Newsletter* (No. 2), Mr Taizo Ino of the Japanese Clematis Society rather takes me to task for writing, in my last book, that the original *C. patens* had 'vanished, as a wilding, into thin air like the Cheshire Cat leaving only its grin behind'. It has been reported in the wild, he says, throughout much of Japan, but confirms that it is a very variable species. He also attributes 'Wada's Primrose' to a wild 'yellow *patens*-type' introduced to Japan from Manchuria in 1933 (see also p. 42).

Raymond Evison saw and photographed what he believes to be *C. patens* in the wild during a visit to Nikko, Japan, in 1984, an event that he recorded, in some detail, in the same edition of the *Newsletter* (No. 2) later that year.

The two names we meet over and over again in accounts of nineteenth-century hybridizing work are 'Fortunei' and 'Standishii'. As Bean (the 1970 revision of *Trees and Shrubs Hardy in the British Isles*) remarks, whether *C. florida* is to be considered as one of the parents of the garden race of large-flowered clematis depends very much on the status of *C.* 'Fortunei'. It was a semi-double, pure white, scented variety of Japanese gardens introduced by Fortune in the 1830s or 40s, and is usually ascribed to *C. florida* but is placed by Rehder under *C. patens*.

It is still common (though less so than heretofore, thank heavens) for gardeners and nurserymen to distinguish a florida group of double, large-flowered clematis but (Bean again) 'in no sense can these be considered as varieties of *C. florida*. They are hybrids of complex origin some of which may have nothing or little of *C. florida* in their make-up. Thus the parentage of "Belle of Woking" is *C.* (*lanuginosa* × *patens*) × "Fortunei".' 'Belle of Woking' has traditionally been placed in a florida group.

If 'Fortunei' is probably a derivative of *C. patens*, 'Standishii' is in the same boat. Introduced by Fortune in 1861, it was a Japanese garden cultivar and was more used in hybridization than *C. patens* itself, but some authorities confuse the issue by considering 'Standishii' to be a cross between *C. florida* and *C. patens* rather than a selection of *C. patens* pure and simple. It seems doubtful whether we shall ever know the truth for certain.

Anyway, the hybrids typically ascribed to and taking after these chimerical *C. florida* and *C. patens* parents flower early on short shoots from the previous year's wood.

Not so *C. lanuginosa,* our third oriental, which bears its flowers successively on young shoots from spring to autumn. Of the cultivars we know and grow, 'Marie Boisselot' and 'W. E. Gladstone' are typical of the lanuginosa style. Their leaves are often simple, untoothed, un-divided. Their buds, stems and young leaves are woolly and the flower buds are long and pointed, not stumpy and rotund.

It was Robert Fortune's introduction of *C. lanuginosa* in 1850 that really set the ball rolling in the hybridizing field. He was collecting for the RHS in the province of Chekiang, at the most eastern tip of China. In his memorandum he wrote: 'This pretty species was discovered at a place called Tein-tung, near the city of Ningpo. It is there wild on the hill-sides, and generally plants itself in light stony soil near the roots of dwarf shrubs, whose stems furnish it with support as it grows. Before the flowering season arrives it has reached the top of the brushwood, and its fine star-shaped azure blossoms are then seen from a considerable distance rearing themselves proudly above the shrubs to which it had clung for support during its growth. In this state it is most attractive, and well repays anyone who is bold enough to scramble through the brushwood to get a nearer view.' The flowers of *C. lanuginosa* are the largest of any species, though the plant, in the wild, grows only 2m high.

Turning now to the European ingredients, the species of prime importance in breeding was *C. viticella*, from Southern Europe, with purple flowers only 3–5cm across, composed of four obovate (broadest near the tip) sepals, hanging lantern-like and never opening flat. It flowers on its young growth from July till October. It was introduced in the late sixteenth century.

The first recorded clematis cross was made between this and, supposedly (the cross may well have been accidental), *C. integrifolia*, a herbaceous European species. This occurred on the Pineapple Nursery belonging to Mr Henderson at St John's Wood in 1835. *C. integrifolia* was never proved to be the other parent but, judging by results, it seems highly probable. *C. integrifolia*, as its name implies, has simple ovate leaves; it possesses no climbing adaptations and its offspring resembles it in this though its leaves are lobed. It is at present correct to call this hybrid *C. × eriostemon* and Henderson's particular cross *C. × eriostemon* 'Hendersonii'. At other times it has been known variously as *C. hendersonii, C. bergeronii, C. chandleri* and *C. × intermedia. C. integrifolia* 'Hendersonii' is another thing.

In 1858 Messrs Jackman of Woking made a (double) cross using *C. lanuginosa* as the seed parent and the pollen of both *C. × eriostemon* 'Hendersonii' and of *C. viticella* 'Atrorubens'. The famous *C. × jackmanii* was one of the two resulting seedlings, first flowered in 1862 and exhibited in 1863, when it caused a sensation. But does it have *C. integrifolia* blood in it (through *C. × eriostemon*) or is it simply the result of the *C. viticella* clone and *C. lanuginosa*? We cannot be sure, but I incline to the latter possibility. If this is right then *C. integrifolia* has played no part in the subsequent spate of hybrids. *C. × jackmanii* founded a race and became the parent of innumerable crosses.

It is noticeable (and it was a thorn in the flesh of the early hybridists, who wanted the largest possible flowers in all their clematis) that the viticella influence, which contributes deep rich purple, is always associated, in this colour, with a smaller flower. The largest hybrids were obtained by crossing *C. patens* (together with 'Fortunei' and 'Standishii') with *C. lanuginosa*, and they had paler shading. Isaac Anderson-Henry of Edinburgh made the first of these crosses in 1855 and obtained the white-flowered 'Henryi', also 'Lawsoniana', which is pale mauve. Both are still in commerce. Anderson-Henry's stock was subsequently taken over by Lawson, whose name is thus commemorated. 'Lawsoniana's' blooms are up to 23cm across.

From now on the breeding of new clematis was a rage in Britain, France, Belgium and Germany. I shall take the opportunity to mention the raisers when discussing their cultivars individually. A. G. Jackman (founder George Jackman's son) criticized some of the Continental hybrids for having long narrow sepals, but when we consider this character in the twisted segments of a variety such as 'Mme Julia Correvon', those of us who enjoy informality may prefer them as garden plants to others with broad, overlapping sepals and a perfectly flat flower, though the latter would doubtless look better on the show bench.

Many clematis in those days were named either after members of the family of the nursery raising them – 'Mme Julia Correvon', 'Elsa Späth', 'Lucie Lemoine', 'Mme Grangé', 'Henryi', 'Jackmanii' *et al.*; or after titled patrons of the said nurseries. We still go on the same principles but the aristocracy, being no longer rich patrons with hordes of gardeners, have dropped out of the running. Firms' hybridizers and secretaries are now more likely to be honoured. If the name turns out to be Fred Bloggs that's just too bad for the clematis should it want to compete in the glamour stakes. Japanese cultivars such as 'Haku Ookan' and 'Myojo' have a type of name unfamiliar to Western ears and may take some getting used to, although at least they are short and readily pronounceable. Nurserymen like short names and for very practical reasons. Think of all those labels that need to be written. *C.* 'Kosmiczeskaja Melodija' is a Russian hybrid ('Gipsy Queen' × 'Jackmanii Alba') which must be straight out of their worst nightmare. I am sure that a synonym such as 'Omsk' would appear in no time. Better by far to give names that have a global appeal, such as those of planets, stars and gods.

From 1880 or thereabouts, the excitement and the breeding died down, though a spark was engendered by the introduction, rather late on, of the red-flowered *C. texensis*, which was used both here and in France for hybridizing at the turn of the century.

Thereafter, hybridists seem to have run out of ideas, and anyway the epidemic nature of wilt disease put a damper on clematis cultivation. Morel of Lyon gave up nursery work and handed over his collection of seedlings to William Robinson, who exhibited them at the RHS from time to time as did Ernest Markham, Robinson's head gardener, after the latter's death; when Markham died these seedlings passed to Jackman's, who named one of them 'Ernest Markham' and another 'Abundance'. Meantime, Jackman's themselves had continued to bring out novelties such as 'Crimson King', 'Duchess of Sutherland', 'King George V' and 'Lady Betty Balfour'.

From the historical account of the development of clematis cultivars as detailed above, it must be agreed that it is misleading to ascribe any given cultivar to any particular group: florida, patens, jackmanii or whatever. These groups are just one more thing for the amateur to learn and they lead him nowhere. The only excuse for their use is as a guide to pruning. Floridas are pruned this way, viticellas that. It would be far more satisfactory, however, to describe the three pruning methods and then to ascribe them to the various clematis that the nurseryman is offering.

A further reason for the nurseryman breaking clematis down into groups is to facilitate their choice from the printed page. He has two alternatives. Either he can print the lot in one complete alphabetical list and indicate, by a code against each variety, what the basic facts that need to be known about every clematis are: whether it is large-flowered or small, whether it flowers early or late and hence whether it falls into one pruning group or another. The advantage of the complete alphabetical treatment is that it is easy to track down any and every variety the customer wishes to find, but he may find a code tiresome.

The other alternative is to break the clematis down into groups. Thus Treasure's used to break them down into no fewer than ten groups: evergreen and early-flowering species, alpina and macropetala, montana types, early large-flowered cultivars, double and semi-double cultivars, mid-season large-flowered cultivars, jackmanii types, late-flowering species (not all species by any means), herbaceous types (some are woody, however). This made for pleasant browsing for the potential customer and informative reading for those more committed to the genus. Without an index it was tedious, combing ten alphabetical lists to find the cultivar of one's choice. It was also inevitably misleading. I note that Treasure's reverted to a single list, covering four sides of eye-catching orange A4 paper, for their 1987/8 season. Some still cling to the old system of classifying according to parent species, but there is little validity in jackmanii and viticella categories any more than there would be in florida, patens and lanuginosa. Why should 'Ville de Lyon' be classed as jackmanii type, to take one instance? It neither looks nor behaves like *C.* × *jackmanii*, flowering on old wood early as well as on young wood later, and its ancestral links with *C.* × *jackmanii* are remote and weak.

Nurserymen act for the convenience of their customers and themselves, but I do think that between us we succeed in painting a pretty muddy picture.

Since this book was last revised in 1979, renewed interest has been shown in the breeding of clematis – and very welcome it is too.

Overseas raisers are busy again, with Jim Fisk acting as a conduit into the UK; 'Niobe', raised by Wladyslaw Noll in Poland, must be counted the outstanding newcomer, while others such as 'Dr Ruppel' from Argentina have proved popular. 'Allanah', 'Prince Charles' and the rather nice 'Snow Queen' are from Alister Keay in New Zealand.

The Japanese are on the move, but few of their hybrids are being offered in Europe. Likewise the Americans. They are thinking big; in 1984 Arthur H. Steffen Inc. had 140 cultivars on the market and 123 more in test plots for possible future introduction!

The Russians, along with the Latvians and Estonians, are producing new cultivars, carrying on a tradition that I am told arose in the 1920s, although as yet none of their hybrids have reached the UK. Jan Fopma, a Dutch grower, recently showed slides of some Soviet crosses to the International Clematis Society. *C.* × *fargesioides* (*C. fargesii* var. *souliei* × *C. vitalba*) did look like a plant midway between both species but not an improvement on either. *C.* 'Mammut' was altogether different: a well-shaped white, 14cm across, with a bushy mass of reddish-purple stamens like a well-used shaving brush that seemed to take up a third of the area of the flower. Produced by Mr Uno Kivistik in Estonia in 1980, I should certainly like to see that one again, soon.

In the UK, Peveril's in Devon are experimenting. *C. viticella* 'Elvan' is one of theirs. In 1986 they also announced the results of a whole series of texensis crosses with large-flowered hybrids, but these have yet to be marketed and appraised. It is exciting to see some work being carried out with these two species again. We did write to Peveril's to find out more, without success. The Dennys, a dedicated husband and wife team from Preston, are to be credited with 'Sylvia Denny', a good semi-double white. John Treasure has been rather hiding his light under a bushel. 'Royalty', a rich purple double/single, and 'Fireworks', raised by him some time ago, have been introduced only recently. John's *C.* 'Pagoda' has been available for longer and is probably better known.

Raymond Evison must take the credit for the introduction of a great many species during his time at Treasure's, not only for inclusion in the National Collection (at the time of writing still held at Burford House, Tenbury Wells), but sold by that nursery and distributed as seed via the International Clematis Society, which came about mainly through his and his wife Hildegard's efforts.

In that 1979 revision, I asked if anyone was working towards a true white jackmanii. 'John Huxtable' is now commercially available and fits the bill, so I shall cross it off my list to Santa Claus. Unfortunately, I cannot make him redundant yet; we *still* want a large-flowering yellow clematis. I called for it in my last book (and the one before) but it seems as far away as China and 'Moonlight' is hardly the answer. Also, we could do with more late-flowering blue clematis of the 'Perle d'Azur' style. 'Prince Charles' is a start, but not altogether what I had in mind. I would like a red that matches up to the rich, pure colouring of *C. texensis*, but don't intend to hold my breath waiting. One of the difficulties, I am told, is that there aren't any decent clones of texensis around with which one could hybridize. John Treasure has been searching for one for years.

On a different tack, I should like to meet some of the notable but shadowy clematis of the past. The blue form of *C. campaniflora*, *C. coccinea major*, the double *C. recta* 'Flore Pleno' and *C.* × *durandii* 'Pallida' or, better still, the variegated version which Raymond Evison tells me used to exist – or is he just pulling my leg?

Finally, I feel compelled to sound a note of caution. Jim Fisk recently wrote Tom that he thought there were too many similar clematis around, with more on the way. He is right. Some selections of alpina and montana are, after all, only minor variations on a theme. I throw in a reprimand for whoever propagates named clones from seed and continues to give them the parent's name. They are genetically different plants. We have had trouble of this kind with *C. orientalis* 'Bill Mackenzie', *C. montana* 'Tetrarose' and *C. integrifolia* 'Rosea', to name but three. A black mark too for those who use an old cultivar name to describe a cultivar in their possession which they would like to use it on. Most often this is a case of mistaken identity, and we now have many a *C. macropetala* 'Maidwell Hall' and numerous *C. armandii* 'Snowdrift' and 'Apple Blossom'. *C. montana* var. *wilsonii*, as a natural variety, has suffered in the same way. We cannot be sure that it was not introduced as more than one seedling, and possibly later from other sources and expeditions. It may well have been variable from the first.

Chapter 3
MAKING THE MOST OF
CLEMATIS

'Of course I should like a clematis, but I haven't the space; I just wouldn't know where to fit it in.' Nurserymen must have heard customers musing in this vein, time without number. Usually the old cliché: 'My garden's only the size of a pocket handkerchief' is thrown in for good measure. These remarks can mean one of two things: either that the gardener in question does not want to be tempted to buy a clematis anyway and is merely inventing reasons that endorse his disinclination, or else that he really does believe what he is saying. While there is nothing anyone can do about altering the first state of mind, the second almost certainly arises from the misapprehension that a clematis needs a special place to itself. The reverse is, in fact, the truth.

Walls

Of all plants, clematis do *not* relish isolation. (For this reason, the notion of a clematis garden, as proposed by sundry author-enthusiasts, when swept away by their missionary zeal, has always struck me as ludicrously cracked.) They are sociable, flourishing on a modicum of competition; good mixers, enjoying the company of their neighbours. (Now and again they may happen to smother and kill one of these, but 'with no offence in the world'.) Most clematis lovers, however, tend to reserve special areas of wall space for their favourites. I am certainly not going to claim that this never works out successfully, but it does tend to emphasize the worst features of the clematis – an expanse of bare and far from shapely legs and, very often, a congested lump of blossom on top. If the clematis shoots had a fairly bulky shrub like an evergreen ceanothus through which to thread their way, then their blossoms would be distributed naturally and without overcrowding. In this way, too, the climber's own fundamental shapelessness would be wholly absorbed by the host shrub's structure, on which the clematis flowers would appear like jewels adorning a handsome woman's coiffure. So my own reaction to a piece of bare wall is to plant a shrub on it first and, when this has got a good lead in two years' time, say, to follow on with a clematis.

It will be worth considering, at this point, which wall shrubs make good clematis supports, bearing in mind that almost any clematis will flourish on an east, south or west aspect, and a great many on a north wall, too. Climbing roses are superlative hosts – not that they possess any structural merit with which to mitigate the clematis's lack of it. Rather is it a case of each attempting to bury its defects in the other's embraces. When flowering, of course, both overflow with the vitality of high summer, and spill their bounty with an abandon that can transform the walls of even the tritest building for a few weeks. Clematis and rose can be chosen either to give a succession of colour or else to flower simultaneously with the maximum impact. The climbing sports of bush hybrid tea and floribunda roses start to flower in late May, just at the time when the large-flowered clematis which bloom on their old wood are getting into their stride. From the pruning point of view these two types suit each other too, for in neither case is it necessary to be drastic and both can be dealt with together.

Clematis combine well with other climbing shrubs of greater vigour than their own and, in this context, wisteria is outstanding. Magnolias provide a good framework but need several years' lead before the clematis is planted, and the same may be said of *Cotoneaster horizontalis* and *C. microphylla*. Pyracanthas, evergreen ceanothus and escallonias are robust wall plants which clematis will much improve, while for a north aspect we have *Jasminum nudiflorum*, *Forsythia suspensa* and *Camellia japonica* hybrids.

If you have any wall fruit trees from which you despair of ever getting a decent crop – either because the fruit turns out not to be worth eating, or because the birds refuse to allow you any without a struggle that seems not worth making – then you should certainly drape them with clematis.

A gardener's natural reaction to the advice I have been giving here is: 'But won't it (the clematis) hurt my rose (forsythia, jasmine, fruit tree or whatever the supporter is going to be)?' And the answer is yes, probably it will, a little, but not enough to matter. This is a danger of which we should not be too frightened. After all, a satisfying garden is not made by growing every plant in it to perfection. That, rather, is the kind of treatment which vegetables should receive. A little competition between different types of plant will enable us to gain a far more pleasing result than if we grow all our treasures in evenly spaced blocks, beds and rows.

Back to our walls. There is one type of clematis which is sometimes invaluable for filling a large space, say two storeys high by 8 or 10m across, and especially so on a rather dull, draughty north aspect. This is *C. montana* and its near relatives. They are too vigorous to be walled up with any other kind of plant, and can easily curtain the whole face, right down to the ground. But beware lest they get on to the roof and under the tiles, which they will lift recklessly. Of course, it is delightful to *see* a montana flung over a roof – someone else's roof.

Artificial support must be provided for clematis grown straight on to walls, and this will be effected either with a wooden trellis or with wires; with Weldmesh or with Netlon, both light plastic materials easy to handle by oneself and usually of decent sober colouring.

Pole, Pergola and Trellis

One of the simplest ways to grow a clematis of medium vigour effectively is on a pole. This will usually be of chestnut or larch, 4m long, the bottom metre tarred for long lasting underground. Some gardeners like to give their clematis support by wrapping a sleeve of wire netting round the pole, but I consider this unsightly, as parts of it are nearly always visible. I find the homely method of tying the clematis to its support with string at 15cm intervals perfectly effective and quite invisible.

The kind of place to site a clematis on a pole will usually be as an isolated, vertical feature rising from a groundwork of lower-growing shrubs or herbaceous plants. Thus it is one of the finest adjuncts for inclusion in herbaceous, mixed or shrub borders of every kind. Sometimes it will be found that the entire weight of a clematis which is being grown over a shrub is too much for the host. A pole strategically placed can be made to take three parts of the burden, while a few strands of clematis are still allowed to wander over the shrub or other neighbouring plants.

Knowing how sensitive rose growers can be about the slightest encroachment on their darlings' territory by foreign bodies, I make my next suggestion with some diffidence, bowing to right and to left and retreating the while. Why not include a clematis or two on poles among your rose beds? The same rich soil conditions suit both shrubs ideally. I have found this practice especially successful among Hybrid Perpetual and Bourbon roses whose 2m growths are being pegged down. You can tie the odd rose shoot to the pole, for a change, while the clematis will occasionally wander away among the pegged rose shoots. An ideal blending is thus achieved.

If you object to the starkness of your poles in winter and want to spare them unnecessary weathering, it is easy and practicable to remove them in autumn and restore them at, say, the end of March (the exact time depending on whether spring is early or late). But this works only where clematis of the type that flower on the current season's young shoots are being grown.

When a row of poles has been planted by way of a feature, then I would recommend associating clematis with climbing or rambler roses, rather than have them isolated. In this context, late-flowering clematis do combine especially well with rambler-type roses. It is only necessary to be prepared to delay the ramblers' pruning until autumn, rather than do it immediately after flowering as tidy-minded rosarians will prefer. This should be no great hardship. The flowering of each will usually overlap, but the clematis is likely to reach its peak after the rose has spent itself. We are always recommended to grow rambler roses in an open situation where there is plenty of air movement, as this discourages powdery mildew from getting a hold on them. The same argument applies to $C. \times jackmanii$ and a number of other summer–autumn-flowering clematis. Moreover, an anti-mildew spray or drench will be as effective on the one as on the other.

Join a row of poles with wire and you have an open fence or trellis which, when planted up with roses, clematis, honeysuckles, vines and other climbers, will make an admirable screen, taking up far less space, in terms of thickness, than the average hedge. The temptation will be, however, to space the plants along it at equal distances, whereas each clematis will be happier in every way if it can be planted with one of the other climbers, which will then automatically shade its roots. Plant them in the same hole, if you like. The same treatment applies where clematis are being worked into a pergola of, say, roses or wisteria. And always plant the clematis on the shadier side of the rose, or whatever it is.

Trees

The idea of growing a clematis over a tree is one which readily appeals, but you do want to know what you are doing, as there are a number of handy pitfalls.

If the tree is alive and flourishing, what chance has the clematis of battling with a creature so much larger and more overwhelming than itself? Any thought of planting a clematis with an oak or other forest tree should be dismissed out of hand. With crabs, ornamental cherries, any orchard tree, rowans, whitebeams and other trees of similarly moderate proportions, the prospects are brighter. But it is rather easy to forget, when planting near a deciduous tree in the dormant season, that its summer foliage will make life very trying for a plant beneath it which is struggling for light. Rowans do not cast too much shade, but cherries are especially umbrageous, while pears and whitebeams are little better. The tree's roots will also be competing for water and nutrients, so that the soil at the foot of a tree trunk is often as arid as a desert's. Here again trees vary; cherries are outstandingly greedy surface feeders while laburnum and the Judas tree (*Cercis siliquastrum*) tend to put their roots down rather than out. A sound rule of thumb is to assume that a tree's roots will spread outwards at least as far as its branches. How slight are a clematis's chances when planted at the base of a tree trunk, is easily imagined. The fact that the tree itself may be young and newly planted makes little difference. The same problem

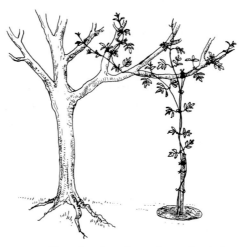

Growing clematis on trees. Plant the clematis on the tree's windward side, if possible, and near the tree's extremities. Lead it with string to the selected branch. Keep a large circle round the plant clear of grass and heavily manured.

obtains but is deferred for a few years.

The solution which is feasible whenever the tree's lower branches reach reasonably close to the ground – within 2 or 3m, let us say – is to plant the clematis underneath a branch near its extremities, and then lead it up by a string to the branch itself. Where there is any choice in the matter, it will be wise to plant on the tree's windward side so that the clematis shoots are blown in the required direction. Once in the tree's branches it can support itself from above, its stem gradually developing into a flexible rope-like liana. The original strand of string can rot at its pleasure.

Many gardeners fancy their tree adorned with the blossoms of a large-flowered clematis. Quite apart from the fact that in many situations and on many trees this would look inappropriate, the main con-sideration here is to choose a clematis which can cope with a tough situation. And the toughest clematis are all small-flowered. But they are none the worse for that and, indeed, small trees offer a fine oppor-tunity for growing a selection of clematis species and becoming acquainted with their great individuality and charm.

It should be added that, as trees are often grown in turf and therefore the clematis will be also, it has here yet another powerful competitor.

However conscientious we are in keeping a large circle around the clematis clear of grass and heavily manured, we shall still need a robust grower.

When a clematis is required to deck a tree which is sickly, its prospects may be altogether more cheerful, with such greatly reduced competition for light, water and nutrients. I feel a bit Jeremiah-like in suggesting that here, too, a hidden danger may be lurking. Why is the tree sickly? And if the tree, why not the clematis also? Mr Basil Furneaux once asked me why an apple tree in my Long Border was dying. 'Old age,' said I, glibly. He shook his head and intimated that there was no such cause. As any fruit grower will have guessed before I say it, the tree was being killed by the parasitic honey fungus (*Armillaria mellea*). This does not confine its attentions to fruit trees. It is practically omnivorous, and if one were to plant a clematis against an infected tree, the former might die of the same cause. And, naturally, the same argument applies to other possible causes of the tree's ill health. It might, for instance, have been starved in some way or else waterlogged.

Still, there is no point in worrying oneself sick by proxy, and it will probably be best to plant the clematis anyway, giving it as generous a start as may be, and thereafter simply keep one's eyes open for trouble so that it can be nipped in the bud if possible. Dead trees should be regarded with the same wariness. They can be an excellent medium for displaying clematis and other climbers, but should not be regarded as permanent fixtures. The more you cover them, the more leverage they offer to the wind. Their top-hamper of smaller branches should be considerably reduced at the start, and if the main trunk does not offer a clematis sufficient anchorage, some strands of wire can be discreetly added as they are required. On the whole, though, my affection for dead trees, as for dead teeth, has waned with the years. Too often they become a source of later trouble; I prefer to have them out.

Shrubs

Of all the practicable and attractive ways in which clematis may be exploited, their association with shrubs offers the greatest scope. Yet to most gardeners this is still an untrodden path. It is only a matter of getting into the right frame of mind. Every mature shrub of reasonable proportions should be regarded from time to time with a questioning: 'How would you look with a clematis growing over you?' Examples of a few that I have found especially suitable hosts are: brooms (especially *Spartium*), cotoneasters, tree paeonies (species), shrub roses, viburnums, berberis, *Hibiscus syriacus*, escallonias, *Cornus* (dogwoods), lilacs, *Senecio* 'Sunshine' and spiraeas.

Shrubs of stiff habit make better supports than those which are whippy and blow about in the wind a lot. It is difficult for a clematis to make initial contact with a plant which will never keep still. One can plant quite close up to a shrub like a rose or a broom, but when it has a greedy, fibrous, surface-feeding root system, like a lilac or a hibiscus, the clematis must be planted well out – perhaps as much as 2m – from the shrub's base, and then led on to its branches. Again, as for trees, plant on the windward side whenever you can.

Tree Stumps and Ground Cover

I am concerned, in this chapter, to discuss how clematis may be used in a general way and to leave consideration of the actual varieties in their special roles till later. Suffice it to say here, then, that certain clematis are good ground coverers, but with the reservation that, being deciduous, they are not properly covering the ground the whole time. The happiest solution is to grow them *among* evergreen ground coverers. Thus an occasional clematis rambling over a groundwork of dwarf heathers (whether summer- or winter-flowering) is very effective indeed. Clematis that flower after midsummer on their young wood are the most useful, here, because they can be cut hard back and tidied up out of the way as soon as they have finished flowering, leaving the field clear for winter ericas.

Large trees, when cut down, often leave a stump perhaps a foot high which is difficult to get rid of. A clematis will do the trick of concealment, but it is nearly always necessary to fix a piece of netting over the top of the stump first.

Ornamental Tubs, Pots and Ali Baba Jars

There is a strong demand for clematis that will enjoy being grown in tubs, pots, urns and the like. Many old country cottages, converted to a higher standard of living, have been made damp-proof only at the expense of a wide concrete footing all round their base, which precludes the planting of wall shrubs and climbers in the ground. The only possible alternative is to grow them in large tubs. Then again the patios of many a town dwelling give little or no scope for planting except in tubs and pots and, indeed, these can often be of such pleasing appearance in themselves that they lend an air to any garden.

In such circumstances, a clematis will be required to grow either up a wall, from the container, or else to flop out of and entwine the container itself – a procedure which is especially effective when the tall Ali Baba type jar is being used. A great many clematis varieties will meet these requirements if they are given extra-specially generous and solicitous cultural treatment. It is, after all, not very pleasant to have your roots cooped up in a pot for life. The strongest (John Innes No. 3) potting compost needs to be used, with additional liquid feeds during the growing season; the largest available container should be chosen so as to allow as much root room as possible, and the container must never be allowed to dry out. It must, of course, have a drainage hole at the bottom. If, after four or five years, the clematis seems to be doing not so well, I should turn it out in the dormant season and completely replace all its soil with a fresh batch. Underwatering is certainly the commonest trouble with container-grown clematis, even when the owner thinks he is being a good and kindly father by administering a few well-chosen daily drops and keeping the soil surface apparently damp. Think in terms of 3 or 4 litres at a daily watering throughout the growing season.

As clematis hate to be hot at the roots, they will be happier with other plants in the same container. For the same reason an east or west aspect will try them less than a south, while for those varieties which will flower freely near or against a north wall (the majority) this position will be better than any.

Chapter 4
THE CHOICE:
GENERAL
CONSIDERATIONS

If you feel bewildered when it comes to making a choice of a few among scores of possible varieties, at least consider how much easier it is with clematis than with roses. The turn-over among roses offered for sale is colossal. Today's novelty is constantly ousting yesterday's. One has scarcely had time to grow fond of a new rose before it is found to have been relegated.

The case with clematis, whose uses are in any case not as varied as the rose's, is far simpler. Many nineteenth-century stalwarts are still going strong, and these are as good a starting point for a collection as any. After all, a variety that has held its own for eighty or a hundred years without falling out of the lists must have cardinal virtues, and so it is with such as *C. × jackmanii,* 'Nelly' and 'Marcel Moser', 'Miss Bateman', 'William Kennett', 'The President' and many more. With the modern varieties, you may be on to a good thing or you may not. Time will tell. You may be attracted by novelty for its own sake and I shan't be sniffy about that either. We are nearly all of us subject to this little weakness, if it is one.

Most of us will strike a balance between enterprise and common sense. It is boring, as you drive through Scotland in summer, to see *C. × jackmanii* and nothing but *C. × jackmanii* being grown wherever there is a clematis to be seen, but at least it is reassuring to know that *C. × jackmanii* is a good doer, and perhaps we can use it with rather more originality than the customary lump hanging out between front window and front door. I have a pole specimen in my Long Border which flowers at the same time as a neighbouring group of *Senecio doria,* this being a glorified ragwort with 2m domes of bright yellow daisies. The contrast is striking, perhaps even crude, but undeniably effective and I only have to murmur 'purple and gold' to myself to feel entirely reassured.

You will find that every list including shrubs of any kind (and even some bulb catalogues do this) will include a few basic clematis that they've imported from Holland. And even if these are wrongly named, as often happens, still, you could do worse than make them your starting point. They'll be basic varieties that you must grow sooner rather than later. The only annoyance, having had your appetite and interest whetted, will be in finding, say, that the illustration of 'Lasurstern' was yoked to an actual plant of 'The President' and that you end up with two of one and none of the other.

Illustrations are a great help where you want to decide on the sort of flower form that appeals to you. Overlapping sepals promote a sense of order, fullness and well-being. Gappy, twisted sepals sound bad, defined baldly like that, but they can promote a no less welcome sense of freedom, informality and movement. You may like blossoms with an eye of stamens in a contrasting colour to form a focal point: the dark stamens in a white bloom as have 'Miss Bateman' and 'Henryi', rather than the pale stamens in a pale bloom – 'Marie Boisselot' and 'Mrs George Jackman'. Or pale stamens in a dark bloom. I certainly prefer the pale central boss in 'Etoile Violette' and 'Lady Betty Balfour' to the nondescript undifferentiation of *C.* × *jackmanii*.

Colour in catalogue illustrations and printed publications must be taken with a large pinch of salt where pink, red and blue are in question. Blues and purples are nearly always better in real life than on the printed page; pinks and reds nearly always worse. The worst deceptions regarding size of flower will be when close-ups of small blooms are illustrated and there is nothing to give scale. This happens most often with *C. tangutica* and *C. orientalis*.

A visit to a flower show can also be helpful, but fewer nurserymen are exhibiting clematis at flower shows than of old because of the great cost of bringing a large number of plants to perfection at the right moment and then of transporting them in mint condition. So it tends to boil down to exhibiting at just one show in the year, notably the Royal Horticultural Society's at Chelsea. Which means, if you get carried away, that you may accumulate a preponderance of early-flowering hybrids in your garden. Plants grown for exhibition have to be raised under cover, of course, to protect them from the vagaries of the weather, and this in itself will result in a different sort of flower – not necessarily better, by any means – from what would be obtained under garden conditions. By and large you get more intense colouring from outside blooms.

Think of the backgrounds that your clematis will have and their positions. Blue clematis will look well with every sort of rose, whether yellow, pink, red or orange. Red clematis are lost against red brick walls, and if there is any pink in your brickwork, certain clematis of a magenta persuasion can look filthy against this. I have a model example of what not to do in a photograph of 'Barbara Dibley', fading in sunlight against such a wall. White clematis against white weather-boarding or Snowcem would be considered chic in some quarters, dotty in others (hear! hear!), but on a north wall or in any shady situation, white or light shades should obviously be used in preference to purples and reds.

Indeed, it can be stated as a general precept that light colours show up best in the general setting of a garden. They will be noticed at a distance, whereas the dark colours such as you find in 'Royal Velours' 'Kermesina' and 'Star of India' are sumptuous and thrilling when you stand up against them with the sun behind you, but on dull days they look glum and at a distance you don't see them at all whatever the light.

Clematis which fade markedly need special consideration. Nothing could be more galling than to struggle to establish a plant only to have the flowers bleach to an achromatic, uninteresting mess almost as they open, the half-expanded blooms tantalizing you with a vision of what could be if only the sun would go away. All the 'Nelly Moser' type fade badly, with the pale-flowered cultivars such as 'Hagley Hybrid', 'Hybrida Sieboldiana' ('Ramona') and 'Silver Moon' not much better. These will fade within forty-eight hours, given midsummer sun on a south-facing wall. You would be far better off with the more emphatic colours of 'Niobe', 'Lady Betty Balfour', 'Ernest Markham', 'The President', etc., or a clematis that still looks presentable when it has bleached. Such is 'Barbara Jackman'.

If you live in the south or east, as far north as Norfolk, you can expect to be successful with most clematis varieties against north or other shady walls. They will bud up without the encouragement of direct sunlight. I should be chary of using *C. orientalis* L & S 13342,

'Huldine', 'Ernest Markham' or 'Lady Betty Balfour', unless left unpruned, then you might have a chance.

In the colder parts of Britain 'Lady Betty Balfour' is useless; others such as 'Ernest Markham' and 'Mme Baron Veillard' are little better. Montana and alpina types would be the safest. Of the large-flowered hybrids I should experiment for a start with 'Comtesse de Bouchaud' and 'Perle d'Azur'. Also 'Nelly Moser', provided the site was sheltered from cutting winds. The larger the bloom the more vulnerable it will be to wind and draught. Remember that walls are very draughty, and that the draughtiest spot in any garden is between the walls of adjacent buildings. Here you get a wind funnel. Only the smallest-flowered clematis can be expected to flourish, and even then I should prefer to back some of the more cheerful ivies.

Try to anticipate with the space you have available the size of a well-grown mature plant. To put an ultra-vigorous montana on to a 2 × 1 metre trellis (and I've seen it tried) would be like taking a tiger cub as a house pet; you and it will very soon fall out. Similarly, to expect the same plant to cover acres of wall when confined at the root to a pot on your patio will involve you in considerable anguish (not to mention watering) later on.

Clematis do not mind the salt-laden air of seaside gardens. They will be far too battered and bruised if placed unsheltered to face the sea, but quite all right if faced inland.

Chapter 5
THE CHOICE:
LARGE-FLOWERED
HYBRIDS

My object now is to make a critical and comparative survey of clematis cultivars available, so that the reader has some guidelines on which to base his choices. The trouble is that I know some of my clematis far better than others, while there are also some that I have never clapped eyes on and so I cannot do them justice. Indeed, it would be foolish to bring them into this chapter at all. I must ask the reader to make the most of the half loaf I am offering. It will be quite a long half, even so, not to say tough in patches.

Starting with the large-flowered hybrids I propose to divide them into colour categories. These divisions will be arbitrary in borderline cases, as between blue and mauve, mauve and purple, purple and red, but never mind.

I must say, at the start, that it is often difficult to decide if a variety as offered in the trade is correctly named or not. 'Mrs Hope', for instance, will appear in one guise in one place, in another guise elsewhere. Who is to say which is correct or whether the original 'Mrs Hope' was not yet a third variety that has gradually been replaced and has now completely disappeared? Nobody can give a final and incontrovertible opinion on these questions of identity, and the best I can do is to weigh the balance of probabilities.

These large-flowered cultivars are very poorly served in their naming. And the situation appears to be as bad now as ever in the past, with Tom Snooks, Dick Drab and Harry Brown being celebrated right, left and centre. Mr Fisk seems to encourage the practice when he writes, apropos of raising a winner from seed, of 'fame for the raiser, if he or she decides to name it after themselves'. This is an understandable piece of vanity but should be resisted. If these breeders would leave us a record of the circumstances attending the origin of their new cultivars, it would be far more to the point. But a name can tilt the balance between the successful promotion of a good plant and its relegation. Percy Lake, Percy Picton and Percy Robinson are (or were) probably all three excellent fellows (the second, certainly, I can vouch) but their names mean nothing to 99 per cent of the clematis-buying public. Neither do they evoke any sort of mood that can be associated with a flower. Compare them with names like 'Moonlight', 'Gipsy Queen', 'Aurora', 'Blue Gem', 'Victoria' – all of them short, easily memorized, easily spoken, in sum: attractive. If clematis namers would try to be a little less indulgent to their relations and friends, a little more imaginative, it would be a great help all round.

───── 'BLUE' CLEMATIS ─────

'Ascotiensis'	'Lady Northcliffe'
'Beauty of Worcester'	'Lasurstern'
'Countess of Lovelace'	'Lord Nevill'
C. × *durandii*	'Mrs Cholmondeley'
'Elsa Späth'	'Mrs Hope'
'Etoile de Paris'	'Mrs P. B. Truax'
'General Sikorski'	'Perle d'Azur'
'H. F. Young'	'William Kennett'

'Blue' has to stand in quotes because even the bluest clematis has a strong tinge (or taint) of mauve in it, but we shall always continue to speak of blue clematis in describing those here listed. The cynic will say that this is a nurseryman's confidence trick, as blue is the best selling colour of all. Rather is it a matter of comparison. There are so many clematis that come indisputably into the mauve-lavender-lilac bracket, that when they do verge on blue it becomes necessary to call them this.

Blue is the most exciting clematis colour. No other hardy climber carries huge blue flowers, and in this it is completely differentiated from climbing roses.

The deeper blue shades are more popular than the light. Here our choice lies between four old and well-tried varieties: 'Lasurstern', 'Lady

Northcliffe', 'Lord Nevill' and 'Beauty of Worcester'. 'Lasurstern' heads them. Blooms carried on the old wood are 15 to 18cm wide. Each has seven or eight sepals, broadly overlapping towards the axis of the flower but tapering to fine points, their margins gently undulating. The creamy-white stamens make a prominent central boss. On a healthy, carefully trained specimen I have had over 200 blooms open at a time and covering an area 1·5m high by 5m across. Really you should try and work it in with a double red climbing rose like 'Guinée' or 'Etoile de Hollande', having the same flowering season.

'Lady Northcliffe' has much the same colouring and is as good in its way, though I have seen it afflicted with mildew. The flower is quite a different shape, without fine tapering of the sepals and unwaved, but the bloom as a whole has a classic purity of outline that is characteristic and always recognizable once recognized. This clematis was raised by Jackman's from a cross between 'Beauty of Worcester' and 'Otto Froebel'. It received the AM when shown by Jackman's, 29 May 1906.

'General Sikorski' was raised in Poland by Wladyslaw Noll and introduced by Fisk's in 1980. It is a blue with more than a hint of mauve (they describe it as 'mid-blue'), with a more pronounced reddish blotch at the base of each sepal. These overlap just sufficiently to make for a full flower without it looking congested. It is a strong and quite compact grower, which are both points in its favour, and it flowers willingly on both new and old wood. A good new introduction.

There is a rough readiness about 'Lord Nevill' (introduced in 1878) that is far from aristocratic, I should have thought. The puckered surface of the sepals suggests a wind ruffled pond. The purple flower centre is even darker than the sepals. In a way I love 'Beauty of Worcester' better than any of the foregoing, but it does not grow as strongly as one would wish. Indeed it is so free-flowering that three-quarters of its effort seems to be channelled in this direction, leaving too little for the important business of building a strong framework. The early flowers are double, rich and intense; the later, by contrast, are so well shaped that you would never guess that doubleness could be an aspect of the plant's personality. The centre of these single flowers is very white. No wonder its marvellous qualities have been much used in hybridization. 'Beauty of Worcester' was raised by Messrs R. Smith and Co. of Worcester – I should guess in the 1890s. Mr Pennell used it in the cross that produced 'Vyvyan Pennell'.

Turning to the lighter blues, my highest award (not counting the later-flowering 'Perle d'Azur') would go to 'Mrs Hope' – a huge satiny bloom with a full and comfortable look that it owes to overlapping sepals. And the reddish-purple eye is an excellent feature. It flowers early and late, and the plant I have been describing makes a speciality of very long, acute-angled, simple leaves. But there are other 'Mrs Hope's around (Treasure's is not the same as mine) and their sepals are more rough in texture. 'Mrs Hope' came out in 1875 (her maiden name is not recorded): 'William Kennett' is often mixed up too but should not be, as it is a distinct and easily recognized clematis. One of the old brigade but still robust and hearty. Rough-textured sepals led to Moore and Jackman's comment of its 'lacking in refinement' and we can agree, but from May till autumn 'William Kennett' is seldom out of flower.

I think I should bring in 'H. F. Young' here. It is paler than 'Lady Northcliffe' but has the same sort of smooth-textured refinement and a good deal more vigour. The photograph of a sumptuous specimen of 'H. F. Young' features on the jacket of Jim Fisk's *The Queen of Climbers*. 'Mrs P. B. Truax', always described in catalogues (and aptly so) as periwinkle blue, comes into the same colour bracket. It is one of the earliest large-flowered hybrids in bloom and is a good doer, but has a short season with nothing in hand for later on.

Apart from its colouring this is in complete contrast to 'Mrs Cholmondeley', a clematis that simply can't stop flowering and yet has enough vigour to get it places. From tip to tip the early flowers are very large, but because of narrow sepals and large gaps between them towards the base, one's impression is not so much of size as of prodigality. Any massed effect in the garden should have a complementary feature to offset it, and here I should like the climbing sport of an old Hybrid Tea rose, 'Caroline Testout' or, even better, 'Lady Waterlow'. Both are pink.

By way of a light blue double-flowered clematis, full of sepals and of great depth, 'Countess of Lovelace'. It can be breathtaking but is not necessarily an easy variety. When suited, it still (after a century in cultivation) has plenty of vigour but may take longer than most to settle down and you should give it the works as to food and drink.

We now come to two cultivars that are not quite as blue as those so far discussed. 'Etoile de Paris' is out in my garden in mid-May, and is well named Etoile as its sepals are markedly pointed. The flowers are fairly deep purple-blue on opening but fade a lot. As I have scarlet oriental poppies nearby, I always think of it in this context. My plant, in twenty years, has never given me a moment's anxiety and it flowers with a will, but all on the old wood with little or nothing to follow. 'Elsa Späth' (often spelt Spaeth in English) is one of my greatest favourites. The colour is a strong purplish-blue and the reddish-purple stamens are effective. The texture is smooth, the flowers substantial, but what I specially like about this clematis is the smart cut of its sepals and the way they overlap and yet remain perfectly flat. Also it is vigorous and flowers freely both early and, even more so, late, on its young growth. The good selling name 'Xerxes' is still creeping back as a substitute for this variety. Ever since the original 'Xerxes' went out of cultivation (A. G. Jackman, in his revision notes for Moore and Jackman's book, made early in this century, observed that 'Xerxes' had been superseded), nurserymen have been wanting to use the name.

Many of the blue clematis so far discussed flower well on their young wood, but all are capable of flowering early on short growths from the previous season's wood; some of them do it entirely this way. We now come to two clematis – only two in this colour – that flower entirely on their young wood. 'Ascotiensis' is the earlier of these, out from mid-June. It is a good mid-blue but the flower is a poorish shape. It was introduced in 1871 and Moore and Jackman lay particular stress on its 'large flowers of fine form'. This is strange; they are not even large, compared with other blues. Still, it is pleasing by any standard and indispensable as the only one of its colour and kind. We do badly need some more late-flowering blue clematis, perhaps by crossing 'Perle d'Azur' with 'Elsa Späth' or 'Beauty of Worcester'.

'Perle d'Azur', indeed, quite eclipses 'Ascotiensis', such a paragon is it in so many ways. It was introduced by Morel and described in the *Revue Horticole* of 1885. The parentage was given by Morel, in a lecture to the International Hybridization Conference held at Chiswick in 1899, as *C. lanuginosa coerulea* (♀) × *C. viticella* 'Modesta' (♂). This is close to the cross that seems likely to have produced *C. × jackmanii*. Clearly this clematis is closely akin in habit and flower shape to *C. × jackmanii* and I cannot say that the individual flower starts a song in my heart, but it is well poised and presented and the aggregate of blossom never looks congested. The colour is obviously not sky blue, as often stated, but we may hope to get away with the description if the sky is overcast.

It gives a blue impression, light blue, lots and lots of it. Never did a clematis have a greater will to grow and to flower – over a continuous 2-months-long season from late June to August. You should certainly grow this with your rambler roses. In fact, again and again, I find myself recommending it as the one and obvious choice for many different situations. This is not popular with nurserymen, as 'Perle d'Azur' is not easily propagated in large quantities unless large stocks are held. The internodes are particularly long, the nodes correspondingly few. Demand constantly outstrips supply, this being aggravated by its being scarcely known, let alone propagated, in the Netherlands, so that the trade in this country has no scope for importing. But anyone who has seen the crescent wall at Sissinghurst Castle, Kent, when covered with 'Perle d'Azur' will know why I keep rabbiting on.

We end up with another winner but it is odd man out. *C.* × *durandii* has simple oval leaves and doesn't even climb. It is the unique result of an early cross between *C.* × *jackmanii* and the herbaceous South European species *C. integrifolia*, and was introduced by Durand Frères of Lyon in 1870. Surely it is time that similar crosses were made again. *C.* × *durandii* (sometimes known as 'Integrifolia Durandii', though this, of course, is wrong as it is a hybrid, not a variety of *C. integrifolia*) is semi-herbaceous and will put on 2 m of growth in a season. You can tie it up to achieve height but against a wall it is prone to serious onslaughts by mildew. Better to let it drape itself over a metre-tall shrub in the open. I have it with the grey-leaved *Senecio* 'Sunshine', and the yellow daisies of the latter are pretty with the deep indigo-blue flowers of *C.* × *durandii*. These are of very moderate size: only 10 cm across, deeply grooved with parallel veins like a plantain leaf and having a distinctive button-like eye of pale stamens. The flowers can be abundant on generously fed plants and they cut well. They are unusual as well as beautiful, and flower arrangers have taken them to their hearts. *C.* × *durandii* can get wilt disease, but it is generally an easy clematis to grow anywhere. There is, or used to be, a paler form called 'Pallida', which one would like to meet again.

——— LAVENDER AND MAUVE ———
CLEMATIS

'Beauty of Richmond'	'Mrs Bush'
'Belle Nantaise'	'Prins Hendrik'
'Hybrida Sieboldiana'	'Royalty'
'Joan Picton'	'Silver Moon'
'Lady Caroline Nevill'	'Vyvyan Pennell'
'Lawsoniana'	'W. E. Gladstone'
'Miriam Markham'	'Will Goodwin'

I have tried to collect the bluer shades of mauve here (though some might have gone into the white group), leaving those that verge on pink till later.

'Lawsoniana' is of special interest because it was one of the earliest large-flowered hybrids (see Chapter 2, p. 12). It was recognized as the largest-flowered of all clematis, with blooms 24cm across recorded. In this it has probably not been exceeded. The blooms are rosy lavender on opening, becoming paler and bluer as they expand to their full size. 'Beauty of Richmond' is very similar and 'Mrs Bush' is too, being close to *C. lanuginosa*. Its anthers are buff as against the lavender-shaded ones of 'Lawsoniana'. 'Mrs Bush' flowers best on its old wood but 'Lady Caroline Nevill', although it can carry large, flattish, semi-double flowers off its old wood, often misses out on these in the garden (especially as you get into colder climates) and concentrates on its later, new wood crop. Mr Pennell, gardening in Lincoln, recommended hard pruning so as to cater entirely for this. It is pale bluish mauve; *quite* nice.

'Ramona' comes in here, though a thoroughly mixed up variety. 'Hybrida Sieboldiana' is synonymous with whatever we are calling 'Ramona' (its American name), but I agree with Mr Fisk that 'Ramona' is much to be preferred as a name and especially since *C. florida* 'Sieboldiana' is the correct name for the plant once more generally known as *C. florida bicolor*. I originally received 'Belle Nantaise' from Pennell's in the dim past before the last war, as *C. lanuginosa*. It certainly has close affinities to that species and did me proud for twenty-five years, carrying very large, white-centred, pale lavender flowers mostly on its young wood. I have lost it and cannot say whether the plant now being offered as 'Belle Nantaise' is the same thing. It was a Continental variety introduced by Boisselot in 1887.

Still in the general range of lavender, lanuginosa-orientated clematis that I have been discussing, I shall include 'Miriam Markham'. Like 'Lady Caroline', it is capable of flowering off the old wood but more often does it all on the young in late summer and autumn. The outstanding feature here, however, is that these blooms are double – flattened and of a charming shape. It is exceptional for large double flowers to be borne on long young current season's growth. According to Jackman's old catalogues single blooms are produced on young wood, but this is not my experience. It is a pretty clematis but never carries many blooms at a time.

'Joan Picton' grew healthily in a garden I visited regularly in Northamptonshire. It was coarse and hearty and flowered well. Each bloom was well shaped and moderately large, mauve with a deeper mauve margin and striking reddish-purple anthers. The plant, as a whole, rather lacked distinction.

'Prins Hendrik' (he was Queen Wilhelmina of Holland's husband) is a dramatic show clematis: it has for long been grown under glass as a cut flower on the Continent and is better known there than here. Not only are its flowers very large but so are each of the six sepals, individually. They are strongly veined in the centre and crimped at the margins, but the reason for this powerful variety being so rare in British gardens is that practically no one can make it grow! Don't ask me why.

'W. E. Gladstone', on the other hand, has long been a favourite. It is typically lanuginosa, flowering quite early if you prune it lightly and then continuing non-stop till the autumn. Never a large number of blooms at any one time, but what there are are of the largest – exceedingly well formed, smooth and silky textured (in complete contrast to 'Prins Hendrik') and with a telling eye of dark anthers. Like 'Lady Caroline Nevill' and others, 'W. E. Gladstone' is readily killed to the ground in winter; it then grows strongly from the base and flowers from July onwards. Noble of the Sunningdale Nurseries introduced this clematis; it was awarded a First Class Certificate by the RHS in 1881.

Let us see what the present has to offer. 'Silver Moon' might be roughly described as a smaller-flowered version of 'Belle Nantaise' –

pale lavender, pale stamens and a long flowering season. It is nicely shaped, and the colour of the newly opened flowers is tolerable but they quickly fade to a washed-out mauve.

'Will Goodwin' is one of Mr Walter Pennell's breeding, not unlike 'Prins Hendrik' in appearance with its crimped, pale bluish sepals, but smaller. It is a full looking bloom. I like it.

'Vyvyan Pennell' (Mrs Walter Pennell) is a cross between 'Beauty of Worcester' and 'Daniel Deronda'. It is a double clematis and, like many doubles, has an outer frame of broad guard sepals and an inner rosette of smaller petal-like organs – they are really modified stamens. There was for many years a dramatic specimen on a house in the High Street of Rye (my local town). It used to stop all who passed in gaping admiration (I won't say it actually stopped the traffic also, because this is usually at a standstill anyway). Indeed, the lady who planted it sensibly cashed in on a ravenous demand and bought in plants for re-sale at her back door. But waves of repercussion constantly reached me out here in Northiam, eight miles away. No sooner did visitors start inquiring about a clematis they had seen than I knew what was coming and could help them out of the agonies of explanation. 'Vyvyan Pennell' has great vigour and, on its young growth, carries single blooms in autumn. But its May–June flowering is the thing. You may sometimes be disappointed. 'Vyvyan Pennell' has a tendency to die back in severe winters or in those parts of the country where hard frosts are more common. It is also very precocious, the buds breaking at the turn of the year or even earlier and often being subsequently frosted, the double flowers dying with them.

'Royalty', a double/single raised by John Treasure about five years ago, tends to keep its head down until late spring and its doubles are therefore more reliable. They are similar to 'Vyvyan Pennell' although less congested and, in our opinion, a better colour – deep purple-blue. It was from seed from 'Maureen', and resembles that clematis in being moderately vigorous and quite bushy.

I have transferred to the whites several clematis that I originally intended to discuss here, so will treat them next.

——————— WHITE AND OFF-WHITE ———————
CLEMATIS

'Belle of Woking'   'Lady Londesborough'
'Dawn'   'Marie Boisselot'
'Duchess of Edinburgh'   'Miss Bateman'
'Fair Rosamond'   'Moonlight'
'Henryi'   'Mrs George Jackman'
'Huldine'   'Mrs Oud'
'Jackmanii Alba'   'Snow Queen'
'John Huxtable'   'Sylvia Denny'
                'Wada's Primrose'

Although far weaker than it used to be, there is still a public prejudice against white flowers. But the large white clematis have so powerful an appeal as often to override dissent. And there are and always have been a great number of them.

'Jackmanii Alba' is nothing like *C. × jackmanii*, though it is said to be a cross between this and our old friend the semi-double white 'Fortunei'. And it was made by Charles Noble at Sunningdale. So we find it, not unexpectedly, carrying double flowers early in the season from old wood but they are not a good shape and their white is impure: bluish, like skimmed milk. Later blooms with only four, five or six sepals are more attractive (if not gashed by earwigs). It is an out-standingly vigorous clematis with characteristic pale green foliage. Not in the top flight, but not to be sniffed at either. It received a First Class Certificate in 1883. As already explained, the lower-graded Award of Merit had not then been introduced by the RHS, which is the reason for the high number of FCCs awarded to clematis between 1863 (to *C. × jackmanii*) and 1890, which was anyway the heyday of their development.

'John Huxtable' is a white seedling of 'Comtesse de Bouchaud' and looks exactly that. It was raised some years ago by Rowland Jackman and named after an employee of the firm. It wasn't immediately dis-tributed commercially, which was a pity since it was the valuable late-flowering white for which breeders had striven over the previous century. It flowers on new wood, in the style of *C. × jackmanii*, in

July and August and is good-tempered and free-flowering – far less capricious than 'Huldine', the only other late white currently around.

If you want a double white clematis we still have 'Duchess of Edinburgh'. Raised by Jackman's, it received the FCC in 1876 and was described in Moore and Jackman as being 'very sweet-scented, so much so that a single blossom will perfume a room in which it is shut up'. Well now, I get little or no scent from 'Duchess of Edinburgh', although my sense of smell is sharp. It is also exceedingly prone to greenness where it should be white and, in addition, there is a large area behind the flower where sepal-like leaves or leaf-like sepals abound. This was a characteristic of some of the double whites that pre-dated 'Duchess of Edinburgh' and one wonders whether it is one of these older sorts that we now grow rather than the true cultivar. A virus infection could also have played a part. However this may be, it is still a very popular clematis, especially with flower arrangers, and the individual blooms, which are at their peak in June–July, last a long time as is the way with the doubles. My plant grows through a specimen of *Hydrangea paniculata* 'Praecox' (which flowers in early July) and causes great confusion to visitors.

Established plants seldom throw late blooms on young wood, but this can happen on young plants and the blooms are always double. 'Belle of Woking' is similar in habit and flower form but the colour 'is a delicate but decided tint of bluish-mauve or silver-grey' (Moore and Jackman). No nurseryman wanting to sell it today would dream of describing it as bluish-mauve. Silver-grey every time. The fact that Moore and Jackman felt it necessary to call the colour delicate but decided indicates, of course, that it is delicate and undecided. But it is a pretty clematis. Not as quick to settle down as 'Duchess of Edinburgh'. I don't know how the name 'Belle of Woking' strikes you but I always feel it would be appropriate to a gallant old steam locomotive. Not so a lady who wrote me apropos of a young plant in her garden. ' "Belle of Woking" has leaves no larger than a mouse's ear, but doubtless she will come flaunting down the street with ear-rings jangling.' So much for the doubles. 'Sylvia Denny' is a good semi-double white without any trace of greening. A well-balanced flower with a slight scent, it was produced by the Dennys of Preston from a cross between 'Duchess of Edinburgh' and 'Marie Boisselot'. It too needs time to get going.

There are three clematis whose flowers are remarkably similar in every way except colouring. These are 'Miss Bateman', 'Lady Londesborough' and 'Dawn'. The first two were both bred and introduced by Noble of the Sunningdale Nurseries (alas, his notebooks were all destroyed in the last war) and both received First Class Certificates in May 1869. All their blooms are opened in the May–June period (except in young plants) but they are both stunners, their flowers most beautifully and symmetrically shaped, their paleness offset by the central cushion of reddish-purple (or purplish-red) anthers. 'Lady Londesborough' is coloured much like 'Belle of Woking', a pale lavender-grey that fades to French grey. The young blooms of 'Miss Bateman', and particularly the first of the crop, have a lot of green in them, especially down the midrib. As the flowers age, the green (as in parrot tulips) vanishes but it is an attractive feature in this variety.

'Dawn' is a modern cultivar said to result from a cross between 'Nelly Moser' and 'Moonlight'. It is blush white, fading to white. I doubt if it will long stay the course, as its colouring falls between two stools. It was introduced by Treasure's as 'Aurora', but when they discovered that this name had been given to a double clematis that gained the FCC in 1877, they changed theirs to 'Dawn'. 'Fair Rosamond' is blush white too, but is more often confused with 'Miss Bateman' than any other. Actually they are quite distinct. The stamens in 'Fair Rosamond' are much more prominent, being nearly twice as long. The sepals are narrower and pointed – broad and blunt in 'Miss B'. There is a definite pink flush in 'Fair Rosamond' even in mature blooms, none in 'Miss B'. Greenness in young blooms is common to both cultivars, but 'Fair Rosamond' is scented; 'exceedingly fragrant', say Moore and Jackman, 'the scent being intermediate between that of violets and primroses'. I cannot improve on that comparison – for we can only describe scents by comparing one with another. 'Fair Rosamond' has certainly not lost its scent as 'Duchess of Edinburgh' seems to have done, but you shouldn't expect anything strong, as in species like *C. montana*, *C. rehderiana* and *C. flammula*. It needs to be sniffed out by a searching and sympathetic proboscis.

'Fair Rosamond' was introduced by Jackman and received a First Class Certificate in April 1873 (doubtless a pot-plant grown under glass). All the pale single clematis I have just been discussing are closely akin to *C. patens*, and if you think of them together you do get a connective feeling of similarity in foliage, habit of growth and flowers.

41

So, too, with 'Moonlight' and 'Wada's Primrose'. They cannot be all that far from the wild *C. patens*. The eight narrow, separated sepals, like the sails of a windmill, are common to 'Moonlight' and to the illustration of *C. patens* in Moore and Jackman's book, although this is admittedly a very variable species.

'Moonlight' has had a confusing history. It was introduced more or less simultaneously by Treasure and Fisk in 1968, from stock sent to them from Sweden, where it had been acquired twenty years earlier from Copenhagen. Fisk named it 'Moonlight', Treasure 'Yellow Queen', but the latter acquiesced to the more appropriate name of 'Moonlight'. It has the look of a wilding and is a charming, unsophisticated flower, cream-tinted, fading to white. 'Wada's Primrose' came to Holland from Japan and was called 'Yellow Queen' by the Dutch, although it is a different clematis from the one Treasure's called by this name (a fact that they did not at first appreciate). It is now back to 'Wada's Primrose', and 'Yellow Queen' is no longer a valid name for either cultivar.

'Gillian Blades' is like a white 'Lasurstern' and shares the virtues of that excellent clematis. It is not a pure albino but has a tinge of mauve. 'Snow Queen' is another with mauvish margins to its spring flowers, although those in autumn can be bicolour: pink bars on white, so marked that Jim Fisk finds that his customers sometimes complain that he has consigned them 'Nelly Moser' instead. It was raised by Alister Keay, a New Zealand grower, though not from a deliberate cross. A healthy but compact plant. The flowers comprise six sepals, rippled at the edges with an eye of deep reddish anthers. I thought it the best clematis shown at Chelsea in 1986.

'Henryi' was named after Anderson-Henry, who raised it all that time ago (see p. 12), and it is still one of the most popular large whites. Its pointed sepals compose a star-like flower and it has a dark 'eye'. It has many admirers but is not one of my favourites and has always refused to grow for me. But whereas I find 'Henryi' difficult, 'Mrs Oud' *is* difficult. Yet a most beautiful clematis always described with complete aptitude as milk-white. The stamens are dark and the flower form most satisfying with overlapping sepals. I wish you luck with her and please let me know when she is carrying fifty blooms or more.

The only obvious difference between 'Mrs George Jackman' and 'Marie Boisselot' is in their stamens. Foliage is almost identical and so is the form and size of the pure white flower, but 'Mrs George' has buff-coloured anthers whereas in 'Marie Boisselot' they are such a pale cream as scarcely to be noticed as a feature against white sepals. 'Mrs George Jackman' was going strong by 1873 anyway and 'Marie Boisselot' before the end of the century. Even in those days the latter was known to be synonymous with 'Mme Le Coultre' (A. G. Jackman notes as much in his corrections to Moore and Jackman), and this is still the case. 'Mme Le Coultre' is the name used in Holland, and it is perhaps most often adopted by nurserymen in this country who import a lot of their selling stock from Holland and don't want to have to change the tags on arrival. A story I have heard in Holland is that Marie Boisselot and Mme Le Coultre were one and the same lady under her single and married names!

This is undoubtedly the finest large-flowered white clematis. Because of its broadly overlapping sepals, the flower looks very full. The fact that it has no 'eye' doesn't seem to matter. It is robust and, if it carries its old wood through the winter, starts flowering in early June. You can (and many do) prune it entirely for late flowers. But this is generally pointless as it will anyway flower abundantly on its young growth and continues right into November. It looks very pretty when associated with the Late Dutch honeysuckle, *Lonicera periclymenum* 'Serotina', both being excellent for shady places.

The capricious 'Huldine' was till recently the only fairly large-flowered clematis (its flowers are 10cm across) to crop entirely on its young wood, with a season running from July onwards – when it has a season at all. It is one of the most delightful clematis at its best. The flower and individual sepals are somewhat cupped, white above, mauve underneath but translucent. If you get the light right there is a mar-vellous see-through effect. Perhaps the easiest way is to grow 'Huldine' over the top of a wall that is higher than yourself. Up a pole is very good, too. This is a rampant grower and quite often it fails to flower altogether; it may flower abundantly in one position in your garden, scarcely or not at all in another. Some gardeners think they succeed with it by cutting all young growth back with shears in July, thereby obtaining a crop of blossom in September. It is advisable, both visually and culturally, to give it a sunny place. 'Huldine' received the AM when shown by Ernest Markham for William Robinson in August 1934. It was presumably of Morel's raising, some thirty years earlier. There is a strong probability of texensis blood in it.

43

———— 'RED' CLEMATIS ————

'Crimson King'   'Mme Julia Correvon'
'Duchess of Sutherland' 'Niobe'
'Ernest Markham'   'Rouge Cardinal'
'Jackmanii Rubra'   'Ville de Lyon'
'Mme Edouard André'   'Voluceau'

If you are expecting a pure red large-flowered clematis you will, as with pure blue, be disappointed. More disappointed, in fact, because colour photography and printing tend to make blue clematis look less blue (or a quite impossible blue that no one could believe) in the trade catalogues and encyclopedias from which we make our selections, whereas the reds become pure red by cutting out all the blue in their colouring.

There are not many clematis in this class. Some of the duller shades appeared early on, viz. 'Mme Grangé' (I've placed her among the purples) and 'Mme Edouard André'. The brighter tones came in with 'Ville de Lyon' in their parentage.

'Ville de Lyon' was introduced by Morel in 1899. He crossed *C. pitcheri* with *C. texensis* (syn. *C. coccinea*), something that I had not appreciated until Hugh Thompson, in a note to John Treasure, referred to an article in the *Revue Horticole* for 1893. This contained a colour plate depicting three of the progeny. One of these (to quote Hugh Thompson) 'appeared to have that curious and characteristic "Ville de Lyon" hue which is so difficult to describe'. Morel gave the parentage of 'Ville de Lyon' as *C.* 'Viviand Morel' × *C.* (*coccinea* × *pitcheri*). I was already aware that *C. texensis* was involved, but it is a surprise to see *C. pitcheri* implicated; it can now claim part of the credit for the remarkable clematis that resulted and the dynasty of 'Ville de Lÿon' hybrids that came after, many of which are extant.

It is interesting that the article, by Edouard André, goes on to say that *C.* (*coccinea* × *pitcheri*) tended to be strongly scented of vanilla. Regrettably, this attribute has not survived in its descendants.

Here the trail goes cold. What gave rise to *C.* 'Viviand Morel' we do not know, and Hugh Thompson, tenacious researcher though he was, admits as much in a later letter to John Treasure. He discovered, from other sources, that it was a 'red', though not outstanding. Viviand Morel, the man, incidentally, was Director of the Lyon Botanical

Gardens and may have procured the *C. texensis* for Francisque Morel to make his cross. I do not know whether the two men were related.

'Ville de Lyon' is a remarkable clematis, even today. The flowers are beautifully shaped, their pure carmine red, on opening, effectively set off by cream stamens. As the flower ages, the sepal margins retain a deeper colour than the central area, which becomes a trifle muddy, like strawberries that have gone off. No clematis shows distress signals more readily when starved of food and drink, the oldest leaves dying off in a most unsightly manner. Grown well, however, the owner is amply rewarded and, by dint of light pruning, can enjoy blossom with little intermission from the last days of May till September.

'Mme Julia Correvon' was another of Morel's breeding, its parentage being given as *C. viticella* × 'Ville de Lyon' (in the *Revue Horticole* 1900). It has narrow twisted sepals in a style that was eschewed by British breeders, but it makes a change from the full type of flower with overlapping sepals such as has always been favoured by the exhibitor. The charming informality of 'Mme Julia Correvon' is widely recognized today. This clematis was unobtainable in commerce when I first heard of it at a lecture on National Trust gardens given by Mr Graham Thomas around 1958. He showed a slide of a plant growing at Hidcote Manor. John Treasure and I propagated it and brought it back into circulation. It is a strong grower, flowering on its young wood in one protracted spate from late June, and is a good bright colour.

'Mme Edouard André' was introduced by Baron Veillard's nursery of Orléans in 1893, having been exhibited at Tours in the previous year. It was said to be a cross between *C.* × *jackmanii* and *C. patens*. Edouard André was at this time editor of the *Revue Horticole*. He always signed his articles Ed. André and the clematis has invariably been known in the trade as 'Mme Ed. André'. It is one of those clematis that seems to achieve the perfect balance between vigour and freedom of flowering: that is to say, it will never worry you by growing too little but will never overwhelm you with superabundance of vegetation, because all excess energies are channelled into flowers and yet more flowers – just the one crop, but starting in mid-June and continuing for a good two months. The flowers are prettily shaped with rather pointed sepals, but they are just a little too dark and matt for perfection. They are flattered in a sunny position against a light background.

'Duchess of Sutherland' is the handsomest red clematis with a large, well-formed bloom, its sepals broad and overlapping and of a most agreeable crimson-red colouring. It may sometimes produce a few early double blooms, but most of the crop, as with other red clematis, is borne on the young wood in late summer and is single. This is, alas, not one of the easiest or quickest clematis to establish but if you are prepared to protect it against wilt disease and to manure and water it well, you should certainly grow it.

'Crimson King' is not of the easiest either, but a good colour, again, and a flatter bloom than the last with darker stamens. It was introduced by Jackman's of Woking and received an AM in 1917. 'Ernest Markham' is just about the only large-flowered clematis of Markham's own raising that is going around. It was handed over to Jackman's in a batch, and simply marked 'Red Seedling'. When Markham died not long afterwards, in 1937, Rowland Jackman named the seedling after him. Jackman called it 'the most beautiful of all reds'. I do not agree but it is certainly an excellent clematis, of a strong bright magenta, with broad rough-textured sepals and dusty-looking stamens. It is worth pruning lightly so as to enjoy a small crop of large early blooms, but most come on the young wood, which is freely produced, for this is a vigorous and easy clematis. Sometimes it is shy flowering. Like 'Lady Betty Balfour' it needs plenty of sunshine, but I have seen it flowering well as far north as Newby Hall in Yorkshire.

I think 'Jackmanii Rubra' is far more beautiful, but perhaps my specially soft spot for this clematis is on account of its having grown and flowered so well for me for many years.* I prune it rather lightly and get some double blooms at the start of its season in early June, this running straight into its main crop, which is single – a beautiful shade with quite a lot of blue in it, and the flower has a nice smooth texture. Unfortunately, like so many other desirable clematis, 'Jackmanii Rubra' is permanently in short supply.

'Rouge Cardinal' has a long and continuous flowering season from early June. Like 'Niobe' and 'Mme Edouard André' it is so willing with its flowers that its growth is only moderate, a trait that can be used to advantage in many situations. In spite of this it has a strong constitution. 'Voluceau', as I have seen it at Sissinghurst Castle, is not

* It died soon after I wrote that.

very striking and needs full sunshine to bring life into its rather dim reddish-purple colouring.

Finally to 'Niobe', and at last a new clematis that is a real departure from the usual variations on old themes. Bred by Wladyslaw Noll in Poland and introduced into the UK by Fisk's in 1975, it is a trouble-free, compact grower with flowers that are truly outstanding. A rich velvety texture, they open a shade of red so intense as to be almost black, gently lightening to a warm ruby. Sunshine brings out its colour without disagreeable bleaching. Opinions about its cultivation vary: Fisk's recommend hard pruning, most other growers light trimming and removal of dead wood. Your choice will depend on whether you want an early-flowering season or a late. Either treatment should give equally good results.

PURPLE CLEMATIS

'Daniel Deronda'	'Mme Grangé'
'Edouard Desfossé'	'Maureen'
'Gipsy Queen'	'Percy Picton'
'Guiding Star'	'Sir Garnet Wolseley'
'Horn of Plenty'	'Serenata'
C. × jackmanii	'Star of India'
C. × jackmanii 'Superba'	'The President'
'Lady Betty Balfour'	'Victoria'

Purple clematis derive their deep, rich colouring from *C. viticella*. First among the large-flowered hybrids to have a purple bloom was *C. × jackmanii* itself, introduced in 1863 when it received an FCC. It is still going strong, and anyone who has got as far as reading this book will almost certainly be growing it already. Enough said. I personally grow the cultivar 'Superba' in my own garden. It turned up as a sport on the Jackman nurseries (I imagine in the 80s or 90s, as it did not feature in Moore and Jackman's 1877 publication) and has broader sepals that are also a little darker in colour. Probably an improvement, but it all depends on what you like.

The hand-drawn black-and-white illustration of *C. × jackmanii* in

Moore and Jackman shows a perfectly symmetrical (and idealized) flower with broad, overlapping sepals that is actually far more reminiscent of 'Star of India'.

'Star of India' is very close to *C.* × *jackmanii* but was actually raised by Cripps of Tunbridge Wells and was awarded an FCC in 1867, so it was a very early hybrid. It is not as vigorous as *C.* × *jackmanii* or 'Superba' but has a good constitution (except for a predisposition to mildew which is shared by *C.* × *jackmanii*), and its flower is far more interesting because its purple colouring is relieved and varied by a broad central band of a reddish-plum shade.

'Victoria' is another clematis close to *C.* × *jackmanii* in every way except colouring, and far to be preferred in that department for it is a light shade of purple that becomes still lighter – almost mauve – as it fades, but never mawkish. It is as vigorous as they come and associates well, in the same season, with 'American Pillar', if you must grow that brilliant and harsh rose (I do, twice over, and very handsome it looks in the distance against a dark yew background). 'Victoria' was a Cripps clematis and received an FCC in 1870.

Yet another Cripps product, and yet another to compete (successfully, I hope) for our favours with *C.* × *jackmanii*, is 'Gipsy Queen', introduced in 1877. Its dark, velvety purple colouring is very near to *C.* × *jackmanii* 'Superba' but the flower is a different and, I consider, a more attractive shape, with six finely tapered sepals and a conspicuous boss of dark stamens, lacking in *C.* × *jackmanii*. 'Gipsy Queen' comes in about a month later, being at its best in August, in the south of England. And it is vigorous; I have seen it competing happily and takingly with the white froth of *C. flammula*, which is even more vigorous. Indeed, I photographed this pretty pair on 14 September one year, which is an interestingly late date. Gardeners always have to write their books in the depths of winter, when it is difficult to recollect details of this kind accurately, so it is reassuring to find the evidence on a slide or in a diary or notebook. Another excellent pairing that I photographed in Scotland on 3 September was 'Gipsy Queen' with the flame nasturtium, *Tropaeolum speciosum*, this being pure red.

'Mme Grangé' (FCC 1877) must obviously come next. It is a very distinct clematis because of the way in which its sepals remain incurved along the length of their margins, even when the flower is fully opened. And the undersides of the sepals as thus revealed are woolly, creating a dusky impression. Some blooms eventually open out flat. There is sufficient red in their colouring to qualify this as a red clematis, but 'deep velvety maroon-crimson, becoming purplish with age' (Moore and Jackman) is more accurate. Vita Sackville-West was crazy about this sumptuous clematis when she saw it in my garden, and acquired several plants for Sissinghurst Castle forthwith. It is almost as vigorous as *C. × jackmanii*. 'Mme Grangé' was raised by M. Théophile Grangé, a nurseryman of Orléans, by crossing *C. lanuginosa* and *C. viticella*, the same cross as produced *C. × jackmanii* in England. Indeed, these two clematis are so similar that the likelihood of *C. integrifolia* having also played a part in the parentage of *C. × jackmanii* (see p. 11) seems remote.

'Lady Betty Balfour' is a unique clematis and marvellous when performing as it should, for it can fill the garden with dramatic colour in September and October when few fresh contributions from bold flowers are to be expected. Chrysanthemums, yes, but the vivid blue-purple (a little reddish in the newly opened flower) of 'Lady Betty', with its lively cream eye, is in absolute contrast. And such a grower. Its vigour is prodigious. But it must have sun and warmth, otherwise it fails to flower altogether, so this is not a clematis for gardeners in the midlands or north. I hope it flowered in Edinburgh for Nora King, who, on planting it, wrote me that it was named after her grandmother, 'whose garden near Woking, laid out by Gertrude Jekyll, was the first to make any impression on my consciousness'. Edinburgh has a very peculiar climate, and it cannot be taken for granted that because it is far north a whole range of sun-loving or half-hardy plants will not do. Evergreen ceanothus, for instance, are a great feature of the Lothians, whereas in the West Riding of Yorkshire they are a dead loss.

'Lady Betty Balfour' received an AM when shown by Messrs Jackman of Woking in 1912, and its parentage is interesting: 'Gipsy Queen' × 'Beauty of Worcester'.

We now begin to move away from the influence of *C. viticella* to larger-flowered clematis capable of carrying a considerable (sometimes a main) crop early in the season. 'The President' was one of Charles Noble's raising. (He was given to using the definite article, viz. 'The Czar', 'The Premier', 'The Kelpie' and 'The Kelpie's Bride'.) It does, in fact, have *C. × jackmanii* as one parent, but carries three crops of blossom from late May onwards and is vigorous enough to make sufficient wood to carry its flowers on. The cupped blossom enables you to admire its silvery underside even when the wind is not blowing. This is a good clematis on the blue side of purple, especially as the flowers age. The young foliage is characteristically bronzed and almost identical with the closely related 'Daniel Deronda'.

And it is from the same stable. Noble must have been a George Eliot fan: he named another clematis after her ('dark lavender in colour with a distinct odour of violets'), but it has dropped out of the field. The parents of 'Daniel Deronda' are said to be *C. lanuginosa* and *C. patens*, so it is a little odd that it should so closely resemble 'The President'. The main difference is in the lighter colouring down the centre of each sepal, at all ages, in 'Daniel Deronda', whereas this paleness only appears in old sepals in 'The President'. Also 'Daniel Deronda' is capable of carrying large, flattened, semi-double blooms early in the season. These can be tremendously exciting, but are usually seen on pot-grown specimens grown under glass; seldom in the garden. I don't think this clematis is anywhere near so prolific as 'The President' but it is good all the same and has a long season.

I got 'Guiding Star', I forget from where, some thirty years ago. It flowers on the old wood more than the young, and has a lively flower with pointed sepals and bright bluish-purple colouring. Recent evidence from several growers suggests that this clematis and 'Lilacina Flori- bunda' are one and the same.

'Sir Garnet Wolseley' is an ancient clematis of Jackman's and as good as ever it was: one of the earliest in bloom, mid-May, with moderately large, well-formed, rounded flowers deep lavender or light purple in colour and dark reddish stamens. It is very free and gives a good crop on young wood. 'Edouard Desfossé' has not stood the test of time so well. It was always esteemed as one of the largest flowered of all clematis – light purple with a conspicuous reddish-purple bar. It was introduced by Desfossé, another nurseryman at Orléans, in 1880 and is still listed but has lost something in vigour, I should say.

Turning to more modern varieties, 'Percy Picton' celebrates its raiser and was introduced shortly after the war. It carries the largest-sized flowers, on its old wood, purple on opening but soon becoming lavender and its anthers are dark. It flowers again on the young shoots. Certainly a good clematis. Sometimes I like its shape and sometimes I don't! Really I prefer 'Horn of Plenty' which is of much the same colouring but has a large and somewhat cupped bloom. Its stamens, as in 'Fair Rosamond', are a tremendous feature. I have no idea where this one came from.

'Serenata', from Tag Lundell who was an amateur clematis enthusiast and breeder in Sweden, was introduced by Treasure's. It is a useful two-season purple clematis. The large early blooms on old wood have six sepals; with their pale centre they remind me of the single blooms on 'Beauty of Worcester' but they are light purple, not blue. Later blooms on young wood are mainly four-sepalled.

'Maureen' is a particularly attractive colour, described in the RHS *Journal* at the time it received an AM in 1956 as violet-purple. There is a good deal of red in this violet. It is not a very vigorous clematis and needs little pruning.

PINK AND MAUVE-PINK CLEMATIS

'Comtesse de Bouchaud'   'Margaret Hunt'
'Hagley Hybrid'   'Margot Koster'
'John Warren'   'Miss Crawshay'
'Kathleen Wheeler'   'Mrs Spencer Castle'
'Mme Baron Veillard'   'Proteus'
                'Twilight'

These all have more or less mauve in them. 'Kathleen Wheeler', indeed, starts a good deal darker than mauve; that kindly warming colour known as old rose. The central veins, of a darker, pinker shade, add distinction to a flower with contrasting creamy stamens. Altogether an acquisition. We owe it to Mr Walter Pennell. Another of his clematis that has lately won wide acclaim is 'John Warren'. In the newly opened flower we not only have a pinkish-mauve bar down the centre of each paler sepal but the sepal margins are dark also, and this makes for a most unusual and striking bloom, especially as the anthers are reddish. The overall colour impression is pinkish lilac and early blooms can be very large indeed. They fade none too pleasantly.

'Hagley Hybrid', an immediately post-war clematis of Mr Percy

Picton's, has much the same habit: compact because it flowers so freely
and over such a long period (from mid-June) on its young wood that
there is little time or inclination for making a lot of extension growth.
This is an encouraging clematis for the beginner, causing little worry.
Its six boat-shaped sepals and dark centre compose a charming flower,
and the dusky mauve-pink colouring is pleasing too, in the newly
opened flower, but it does fade rather too far. 'Twilight' is a more
recent product of Mr Picton's and has a well-shaped flower, more
rounded than 'Hagley Hybrid' and with pale stamens against pink-
mauve sepals. The last three all bloom on their young wood from
June–July. And so does 'Margaret Hunt', which I include entirely on
the strength of its picture in Mr Fisk's catalogue. Once again, it is
dusky pink-mauve but I like the cut of its jib. The six sepals are
rhomboid, rather sharply angled and with uniform gaps between them
that actually suit the general symmetry of the flower. How it performs
in the garden and whether it fades badly, I do not know. I must in
honesty record that when I saw it at the Chelsea Flower Show in 1968
I noted that it was 'a miserable thing. A poor, smaller-flowered, dimmer
edition of "Hagley Hybrid".' But then it must have been forced, for
that show. Colour in garden-grown plants is always more intense.
Having read the above in the 1977 edition of the book, the eponymous
Margaret Hunt, raiser of the clematis in her Norwich garden, wrote to
me enclosing a photograph of her plant growing over her garden fence.
Without reservation it looked splendid, prolific beyond probability.

'Mrs Spencer Castle' dates back at least to 1913 when, on 1 July, it
was shown to the RHS but received no award. It is pale mauve-pink
but large-flowered and semi-double when it blooms on the old wood.
Like 'Lady Caroline Nevill', however, this one often misses out on the
early performance and does it all on the young wood. 'Miss Crawshay',
on the other hand, although of much the same colouring – a little
pinker, actually – flowers entirely on its old wood in early summer and
is habitually semi-double, opening wide to reveal pale stamens. Very
charming. Jackman's introduced it around 1873. 'Proteus' is a more
dusky shade of pinkish-lilac. Introduced by Noble and awarded the
RHS First Class Certificate in 1876, it is still a remarkable and outstanding clematis, so full of sepals at its first flowering as to resemble
a rose or, perhaps even more nearly, a double opium poppy. Single
flowers borne on young shoots are meek little things. 'Proteus' would

'Ville de Lyon'

TOP: 'Moonlight'
BOTTOM: *C. × jackmanii* with *Cornus alba* 'Elegantissima'

TOP: 'Nelly Moser'
BOTTOM: 'Comtesse de Bouchaud' prostrate on a lawn

TOP: *C. florida* 'Sieboldiana'
BOTTOM: 'W. E. Gladstone'

TOP: 'Elsa Späth'
BOTTOM: *C. flammula*

TOP: 'Mme Grangé'
BOTTOM LEFT: *C.* × *cartmanii* 'Joe'   BOTTOM RIGHT: *C. forsteri*

TOP: 'Venosa Violacea'
BOTTOM: *C. verticillaris* (syn. *C. occidentalis* var. *occidentalis*) in the wild

'Royalty'

easily find a place in the top six best double, large-flowered clematis.

'Comtesse de Bouchaud' must be accounted the most successful pink moderately-large-flowered clematis ever raised. It is close to *C.* × *jackmanii* in habit and appearance but, like 'Perle d'Azur' with which it associates so well, was raised and introduced by Francisque Morel some time between 1900 and 1906 (when Morel ceased to be interested in clematis). Ernest Markham got an AM for it in July 1936. There is nothing subtle about 'Comtesse de Bouchaud'; its shape is ordinary, its light pinky-mauve colouring might be accounted crude, judged in a vacuum or against an unflattering background, though it fits in excellently with the predominant greens of the average garden setting. But this clematis has a tremendous will to live and luxuriate and carry masses of blossom with the minimum of worry on our part. The season is June–July and there is little left after early August.

'Margot Koster' has a moderately early season, considering that it flowers on its young wood. If you went by the diameter of its flower pinned out like a butterfly on a board, you would class this as a large bloom, large as 'Comtesse de Bouchaud', but its sepals are narrow (after the Continental manner) and twisted and altogether delightfully informal and of a bright rosy colouring.

Finally 'Mme Baron Veillard' (introduced by Baron Veillard of Orléans in 1885), which, although of a lilac-rose colouring that we have met several times already, falls into a class of its own on account of its late flowering season – from the end of August till well into October. Like 'Lady Betty Balfour' it may therefore fail to come to the boil in colder areas. The flowers are well shaped and freely produced. For those of us who can make a go of it, a thoroughly useful and welcome clematis, and vigorous withal.

'NELLY MOSER' TYPE
CLEMATIS

'Barbara Dibley'	'King Edward VII'
'Barbara Jackman'	'King George V'
'Bees Jubilee'	'Lincoln Star'
'Bracebridge Star'	'Marcel Moser'
'Capitaine Thuilleaux'	'Mrs N. Thompson'
'C. W. Dowman'	'Nelly Moser'
'Fairy Queen'	'Sealand Gem'
'Fireworks'	'Wilhelmina Tull'

Many clematis have a deeper or more brightly coloured bar running down the centre of each sepal than is present in the remaining area, but there is a group of which 'Nelly Moser' is typical that makes a speciality of this bicolor effect. These clematis tend to have fairly large flowers and pointed sepals and they all fade notably as the bloom ages. It seems sensible to consider them together.

Because of their fading, which commonly leads to a disappointingly washed-out appearance in the bloom's later stages, 'Nelly Moser' type clematis are frequently recommended for north walls and other shady sites. This is sensible, but they will grow as happily in sun if required to.

'Nelly Moser' was raised in France by Moser, the Versailles nurseryman, and was introduced in 1897. Next to C. × *jackmanii* it is probably the most popular large-flowered hybrid to this day, for it makes a terrific show in its late May to early June season, and it has a robust constitution. The sepals, although long, are rounded at their tips so that the bloom itself is wheel-like, its vivid carmine bars taking on the role of spokes. I have never been able to make out why 'Nelly Moser's' intensity of colouring varies so much, even in the newly opened flower (before fading) and on the same plant in different years and in different parts of the same season. It is a generally observed fact that the intensest colouring, when almost the entire bloom may be carmine, occurs on October or even November blooms carried terminally on the young shoots. Autumn flowers on this cultivar are a welcome extra but they amount to little in terms of display.

'Marcel Moser' came out a year earlier and is very similar in colouring but has a larger bloom and more pointed sepals. Those who grow it successfully will agree that it is superior to 'Nelly'. It doesn't grow so strongly, however,* but that is not to say it is a difficult clematis. Given good soil it is a good doer and remarkably long-lived.

'Dr Ruppel' is Jim Fisk's 1975 introduction from Argentina, where it was raised by the eponymous doctor. Tom quite likes it; I do not. It has proved popular in recent years and is a more emphatic bicolour than most – Fisk's describe it as rose madder with a deep carmine bar. There is a large boss of golden stamens at the centre of the bloom, and the eight overlapping sepals ripple at the margins. The flowers are a little smaller than 'Nelly Moser' and can vary in colour, those appearing in autumn being quite often monochrome. It is moderately vigorous.

'Fairy Queen' was introduced in 1877, twenty years earlier than 'Nelly Moser', and was one of Cripps's clematis; it was one of the first I owned as a boy and was a great thrill on account of its enormous blooms, 23 cm across. They are pale flesh with a pink bar and flower in June. 'King Edward VII' seems to have undergone a sea change. Jackman's bred it by back-crossing one of their Wokingensis hybrids (which take after their *C. texensis* parent) with a large-flowered hybrid. A. G. Jackman, in a note for the revised version of Moore and Jackman that never came to fruition, observes that in this cross the campanulate shape completely disappeared. 'King Edward VII' was the result of crossing 'Fairy Queen' and 'Sir Trevor Lawrence' and the latter is the result of a cross between *C. texensis* (then known as *C. coccinea*) and 'Star of India'. But Jackman describes 'King Edward VII' as pucy-violet, which is hard to square with what we have now: a betwixt and between sort of colouring with a pale crimson bar but otherwise indefinite colouring, although the flower has substance.

'King George V' is another strange cultivar. It has always had a reputation for shy flowering but has never been dropped by the nurserymen. Blooms are more apt to be carried late in the season on young wood than in others of this class and they have a singular beauty, especially as seen on young stock on the nursery! The dusky background to the bright pink bar conveys an extraordinarily sensuous, tactile quality. I have never seen or heard of a specimen giving a good account of itself in the garden, though it is hardy enough.

---

* This was a recorded fact even in its earliest days. See *Revue Horticole* 1898, p. 303.

A large proportion of the developments in modern clematis have been along the 'Nelly Moser' trail. 'Bees Jubilee' is brighter and better if it thrives for you. It does not lack vigour but is markedly prone to wilt disease. From the same house of Bees came 'Sealand Gem', a mauve without the pink tints of 'Nelly Moser' to quote Rowland Jackman's description in a lecture to the RHS. A strong clematis but somewhat lacking in character.

Then the two Barbaras. 'Barbara Dibley' has the big flower and long pointed sepals of 'Marcel Moser'. It is a brilliant and most exciting colour in the newly opened bloom, and would really qualify for the red group. Its description as pansy-violet has always appealed to the public (as periwinkle-blue for 'Mrs P. B. Truax' and milk-white for 'Mrs Oud'). This colouring certainly is to be found in pansies but so are many other colours, and pansy-violet means nothing special to me. Still, that's a personal matter. The centre of the flower is dark, also, but the whole creation fades lamentably to a kind of wasted raddled mauve. Actually even this can strike one as rather beautiful in an off-beat way (like 'blue' roses). If pruned hard it flowers well on its young wood, but that is not normal treatment.

'Barbara Jackman' is one of the best of this bunch. The background to the carmine bar is far brighter and bluer than in any of the others and it can afford to fade. Old blooms retain their charm and mix well with the young. No need, therefore, to protect this one from the sun. It grows well.

'Capitaine Thuilleaux' is one of Fisk's introductions from France but is often listed under 'Capitan Thuilleaux' and 'Souvenir de Capitan Thuilleaux' (see p. 202). The flower's bright pinky carmine bar is exceptionally broad; the pale margins are correspondingly narrow. The flower form is as in 'Nelly Moser' and the dark anthers likewise. A striking clematis but one gets the feeling, after a time, that there are too many in this category with a close family resemblance.

Pennell's have introduced those that follow, and 'Lincoln Star' (someone said it sounds like a race-horse) impresses me as the best of their bunch. As with 'Nelly Moser' the flower colouring can vary a great deal, but in the garden, in early summer, it is a vivid raspberry-pink that suffuses nine-tenths of the sepal area. The blooms are of only moderate size and truly star-like, with sharp points. It is not an over-vigorous clematis, but is happy once it has settled down. 'Bracebridge

Star' has a similar flower form and is elegant. I don't hold gaps between the sepals against it, but the lavender colouring is not very special.

'C. W. Dowman' is the pinkest clematis in this group that we have seen yet. If only it would hold the colouring of the newly opened flower it would be wonderful, but it quickly fades to non-description. 'Mrs N. Thompson' is (unlike its namesake, I hope) a real eye-catcher that sells on sight. The carmine bar is set dramatically against a deep violet base. The flowers are all sorts of shapes and not at all well formed when, as often happens, there are only four sepals. Although this clematis can grow strongly when established, it is a slow starter. Jim Fisk describes 'Wilhelmina Tull' as an improved 'Mrs N. Thompson' but does not elaborate.

'Fireworks' from John Treasure has rather thin, long sepals with a kink towards their tips which, John tells us, gives an impression of a lighted catherine wheel. He says that its colour is similar to 'Mrs N. Thompson' and it was raised at Burford in the early 80s, a seedling of 'Maureen'.

Chapter 6
THE CHOICE:
SPECIES AND SMALL-
FLOWERED HYBRIDS

FOR CONSERVATORY
AND SHELTERED WALLS

C. afoliata	*C. florida* cv's
C. armandii	*C. napaulensis*
C. cirrhosa var. *balearica*	*C. paniculata* 'Lobata'
	C. phlebantha

It is mid-February as I write. Only in the dead season is there time for this sort of major operation. I am surrounded by books, notes made over the past ten years, trade catalogues, dogs in various attitudes of abandon before the fire – but many's the time when I should like to be able to nip out into the garden and verify some point or other that I am wanting to describe. To no avail. But there is one co-operative member of the genus that is now at the top of its form, *Clematis cirrhosa* var. *balearica*, better known as *C. calycina*. I picked a trail from the garden yesterday afternoon, boiled its stem for 40 seconds, immersed it in water for an hour and it stands radiantly before me. There are twenty-four buds and flowers on it. They hang like green-mouthed bells, but when you look into one you see the reddish-brown freckles (against a pale green background) that are typical of the clone of this plant commonly grown in Britain. They look like mottled eggs, my gardener remarked, and two days ago a visitor (only keen visitors come near us at this season) likened these flowers to a hellebore's, as well he might, for they are close cousins.

It isn't just the flowers, in their axillary clusters of one, two and three, that one admires, but the boldly toothed evergreen foliage, lightly bronzed at this winter season. Many evergreens take on bronzed or purplish tints in the cold weather. It's the vegetable equivalent of a bottle nose.

And then there is its scent. Seldom is it warm enough in the garden for this to make an impact (whereas *Mahonia japonica* knocks you backwards in the current mild weather), but as soon as you bring the flowers into a living room they release their citrus fragrance, which is both astringent and sweet. Or they will fill a conservatory with their scent, provided they do not have to compete with *Jasminum polyanthum*.

My *C. cirrhosa* var. *balearica* grows on an 8ft wall and is planted on its north side, but most of its growth is on the wall top, where there is most light, and if we get a hard frost its blossom will all be ruined. I should do better to grow it on a really high wall from whose protection it could not escape, or in some other sheltered corner, perhaps under an overhanging eave, where frost is a rare visitor. The plant is hardy through most of Britain – a point that was discussed by the RHS Committee which gave this clematis an AM in January 1974 when Graham Thomas showed it. (Odd that such an old favourite should be thus honoured so late in the day, but the putting up of plants for award is a hit-and-miss affair, and probably makes little difference to the popularity of a stalwart of this kind one way or the other, since it is already widely acclaimed.)

Where conservatory space is available it will be well deserved, and *C. cirrhosa* var. *balearica* will give you the greatest pleasure. It will then, at a minumum temperature of, say, 6 degrees C, start flowering in late autumn. As soon as flowering has ended, the plant should be pruned hard back to a metre or so from its roots, for this is a vigorous species and will otherwise occupy far too much space. The same treatment applies to several other clematis deserving of conservatory space and benefiting from it either because of their hardiness problems, their frailty or simply because they can acquit themselves so much better if not harassed by our stormy climate.

The species itself varies widely, and can be found in virtually every country bordering the Mediterranean and its many islands. A great variety of forms exists in respect of the fineness of the leaves, size and markings on the flowers and hardiness. Attempts have been made to classify the species. None that we are aware of have succeeded. Indeed, new forms are still being introduced from the wild. *C. cirrhosa* var. *balearica* is generally accepted as having a finely cut leaf but I would suggest that the distinction is purely arbitrary, for I still have the clone that won the 1974 AM and its leaves are lobed. This has no bearing on the size of flower or its markings, for we have seen large, well-speckled flowers on plants with both types of leaf. It has been said that var. *balearica* is hardier than *C. cirrhosa* and vice versa. With such a widely distributed species, this mix-up is par for the course. Some forms will be hardier than others, depending on the harshness of the climate of the region from which they originated.

C. napaulensis is closely related to *C. cirrhosa*, is rather less hardy, and would similarly benefit from greenhouse protection. My knowledge of it is confined to one large specimen on a high garden wall at Malahide Castle, Co. Dublin, in full flower at the time of my visit one February, though often blooming in January even in the open and in the late autumn under glass. It was absolutely smothered in blossom and quite a spectacle, notwithstanding its modest colouring. The notable features of this species, in its season, are the crowded axillary flower clusters and the purple colouring of the long and conspicuously extruded stamens. There is no scent and the evergreen foliage, at its best, is moderately unattractive. This species received an AM as a cool-greenhouse plant when exhibited by the Director of Kew on 22 October 1957.

C. phlebantha also comes from Nepal but was discovered by Polunin, Sykes and Williams as recently as 1952. It grows in the west of that country on hot dry cliffs at 2,500 to 3,500 metres and is sometimes of a trailing habit, at others makes an upright bush, only 0·6m high. Even as a trailer its branches, in the wild, are only 1·5m long. It is non-climbing but needs, in this country, a warm wall for protection and support. At first it was grown at Wisley under glass, but it seems to be pretty hardy. Well worth growing, as much for its leaves as for its flowers. The leaves are pinnate and hairy, so much so on their under-sides as to make them silvery here. There are usually seven pinnae and these are boldly lobed. The leaf surface is heavily grooved with veins, and the conspicuousness of the flowers' veining suggested the specific epithet. *Phlebantha* means grooved flower. These flowers usually have six sepals and are white, of an excellent shape (reminding me of *C. fargesii*), 3–4cm across and therefore bold enough to be effective.

The specimen I saw at Wakehurst, in Sussex, was four or five years old and clearly showed how differently a species can be expected to behave when it comes into the flesh-pots of English garden life. This specimen was 2·5m high on a wall of that height, and 5m across. Its freedom of flowering doubtless depends on how much ripening its old wood has received and how much of this it has brought through the winter. By early August only a few blooms remained, but the main flowering season is in June–July.

C. phlebantha received the AM on 6 August 1968.

I now come to a clematis that has suffered a most unfortunate, because confusing, name change. For years, right back into the last century, it has been grown and greatly esteemed as *C. indivisa lobata*. But the latest edition of Bean tells us inexorably that we must call it *C. paniculata* 'Lobata' whereas the clematis we knew as *C. paniculata* is correctly *C. maximowicziana*. What a bore.

Anyway, this clematis is one of the finest New Zealanders and like all of that group it is dioecious. Its evergreen foliage is pleasantly unobtrusive and the blossom is magically pure. I grew it on a warm wall for several years and it flowered in April. Then came a hard winter ... But under glass we could enjoy it without meteorological palpitations, in March at the latest. In a cultivar named 'Stead's Variety', *C. indivisa lobata* won an FCC when shown to the RHS by Lord Aberconway on 1 May 1934. It was described as tender, then, but must have been grown outside to be flowering so late.

Others from New Zealand could be worth trying in a sheltered spot. Dr Jack Elliott grows several in Kent, which is not noted for a gentle climate. He suggests that *C. petriei* (syn. *C. forsteri* subsp. *petriei*) is fairly hardy. The flowers, which are produced in profusion in April, are 4·5cm across (though the species is variable) and comprise green sepals, overlapping and pointed at the tips. They have a strong citrus scent and are followed, in female plants, by fluffy seed heads. Male flowers are the showier, and sexed plants are obtainable from some specialist nurseries.

C. × *cartmanii* 'Joe' is highly desirable, but generally needs to be given the protection of a cold greenhouse. Jack Elliott has it in a pot, growing rather like an alpine, and it couldn't look better. It was smothered in flower on 28 March, each one held erect above the low-growing, evergreen foliage. The individual flowers are well formed, about 3·5cm across with six or seven sepals, rounded at the tips and greenish towards the centre. The yellow stamens form a compact eye which adds to the overall neatness of the flower.

And so we come to the famous *C. armandii*, always in great demand, always in short supply. Its popularity is founded partly on solid virtues, partly on sentiment – genuine but unjustified.

C. armandii was introduced at the comparatively late date of 1900 by E. H. Wilson, and is a native of central and western China. It is hardier than is even now generally supposed and can thrive, on a warm wall, in such an inhospitable spot as the City of Birmingham. A sunny wall is advisable, in general, but if there is a large shrub, as it might be a myrtle or a ceanothus, projecting from the wall over which the clematis can fling some of its vines, it will look all the better.

The main reason for armandii's popularity is because it is evergreen, and gardening journalists are always cracking up its splendid, bold, handsome evergreen foliage. *C. cirrhosa* is popular for the same reason and more justly, I think, but then *C. armandii* does have the edge over the latter in the department of flower power. The flowers are up to 5cm across and fairly packed into each axil of the previous season's long trails. They can make a most impressive display. On the other hand they can fail to make any display. The reasons for this are not always clear. That the swelling buds can be damaged by early spring frosts is an undeniably frequent cause of failure. And, although one so often reads that only the large-flowered clematis hybrids are afflicted by wilt disease, *C. armandii* is actually one of the commonest victims. It is not so often that the whole plant is cut off near ground level (though this does happen) but more that parts of the plant, branches and vines, are hit and die in a most unsightly manner on any part of the plant. *C. armandii* is also, incidentally, the one and only clematis I have known to be attacked by sap-sucking scale insects.

But there is another ticklish question. Some armandii stock going around carries miserably small flowers as a matter of genetical inevitability. Why don't the nurserymen propagate from stocks that are known to flower well, you may wonder? Well, some of the best flowerers are the most difficult to propagate. Jackman's used to have a magnificent clone called 'Snowdrift' (the fact that you may come across 'Snowdrift' in a trade catalogue does not necessarily mean – probably won't mean – that it is the same clone as the one I am writing about). But I have always found it quite swinish to propagate from cuttings. Whether for this reason or for another (the firm has dropped one-third of the clematis varieties offered since it changed hands), 'Snowdrift' is no longer stocked by Jackman's. As I said at the start, *C. armandii* is always in excess demand and nurserymen are continually obliged to buy stock in, so it is difficult, often impossible, to know whether they are trading in a good strain or not. The plants will have been retailed long before they have flowered. I sometimes wonder if imported plants are not often mere seedlings, so mixed are they in appearance. It is well known that some of the earlier types from Wilson's seed produced handsome foliage but small, greenish-white flowers. I should be wrong to paint too black a picture, though, because there is a lot of good stock in this country, too, and much of it is propagated here from parents known to be good performers.

The most popular clone of *C. armandii* is 'Apple Blossom'. It is a wonderful selling name. 'Snowdrift' is good but definitely shuddery. 'Apple Blossom' spells April, the lilt of spring, balm in the air, bees, butterflies, bird song ... tra! la! la! and tirralirra! I hate to disappoint, but the glow on 'Apple Blossom's' cheek is of the ruddy, chapped variety, most noticeable on the opening buds, disappearing (mercifully) on the fully expanded bloom, but a blush may return shortly before sepal-fall as is so often the way with ageing white flowers. 'Apple Blossom' received an AM on 9 March 1926 and an FCC on 7 April 1936. These exhibits will have been from plants grown under glass and they point the moral. 'Apple Blossom' grown under protection really does deserve its name. A pink blush is retained in the expanded bloom, the flowers are soft and lovable, unbuffeted by harsh spring weather and the *scent* ... I haven't mentioned *C. armandii*'s scent up to now but it is very strongly of vanilla, even in the garden. In the conservatory it is overwhelming.

The leaves of *C. armandii* are very large, smooth, hard to the touch. They rattle in the wind, disconsolately. You may say that they prattle in the wind, companionably, but I know better. All that noise means that the foliage is being bruised, and it does become exceedingly dowdy long before winter's end. The more reason for choosing a sheltered site. But the leaves are a marvel, I do admit, when very young and a beautiful light bronze colour.

The revised Bean recommends pruning *C. armandii* as little as possible but I think this is wrong. It is extremely vigorous, will make 4 or 5 m of growth in a season, and is not in the least reluctant to break from older wood. I knew a specimen grown on a cottage wall where space was limited and its owner pruned severely every year after flowering. The results were spectacular. Unpruned specimens tend to accumulate a lot of unsightly debris.

An early spring-flowering species that will appeal to the specialist is the curious and fascinating rush-stemmed clematis, *C. afoliata*, from New Zealand. Its green stems perform the function of leaves and the latter are reduced to a mere tendril. There was a photograph of a vast specimen in Ernest Markham's book (published 1935) and he wrote that it had long since reached the top of an 8ft-high wall. I have never seen one like that. It is worth risking in a sheltered place outside but would be safer under glass. And perhaps there one might catch a whiff of the 'most pleasant daphne odour' Markham describes. My clone has yielded my eager olfactories no scent at all. This clematis received an AM when exhibited by Miss Willmott of Great Warley, Essex, on 11 May 1916. It was described as 'a hardy species suitable for walls'.

Because I don't know where else to fit them in I shall discuss the two extant cultivars of *C. florida* here. They are not early flowerers, like the foregoing, but they are good conservatory plants. Generally, however, they are grown on warm walls in the gardens. They have a frail and ancient look that does not bely them. They seem to have stepped out of an old hand-painted illustration and it is a compliment to have them with us still, in the flesh.

C. florida 'Sieboldiana', more often met with as *C. florida bicolor*, but also as *C. sieboldii*, was introduced in 1836, having originally been a variety cultivated in Japanese gardens. It is a wispy plant but may often surprise its anxious owner by growing and flowering with a will and apparently uninhibited by its ancient lineage. On other occasions it gives up the ghost without a struggle. Often likened to and mistaken for a passion flower, its structure is really far less intricate, but it is a far more effective flower in the garden. Its six regular and well-formed white sepals frame the central feature, which is a boss of crowded staminodes or petal-like stamens; bright purple with paler flecks. As with so many double flowers, they are long-lasting; indeed, the centre of the flower lasts for quite a time after the sepals have been shed. The plant makes a good deal of young growth before starting to flower on it, in late June or early July, its season extending well into August. If you want something unusual and beautiful, here's your challenge, but don't ask for my shoulder to cry on if you fail.

Its team-mate is *C. florida* 'Alba Plena' (syn. *C. florida* 'Plena'). It was illustrated in *Curtis's Bot. Mag.* in 1805 and is exactly the same today. It is the prototype to which the odd shoot of 'Sieboldiana' occasionally reverts. Its habit is identical with *C. florida* 'Sieboldiana's', but the central rosette of petal-like stamens is considerably enlarged and the overall colouring is greenish white. In 1975 I wrote that I hoped nurserymen would do all they could to keep it going, for its extinction would be as lamentable as the dodo's, and there would be less excuse: it wouldn't make a meal. That plea is no longer relevant, for it now appears in many specialist growers' lists. Its propagation can be frustrating; 'Alba Plena' has an unpredictable habit of reverting to the more common 'Sieboldiana' (and, as mentioned earlier, vice versa). Often only one shoot reverts.

THE ATRAGENES

C. alpina	*C. alpina* 'Ruby'
C. alpina 'Burford White'	*C. alpina sibirica*
C. alpina 'Columbine'	*C. alpina sibirica* 'White Moth'
C. alpina 'Frances Rivis'	*C. macropetala*
C. alpina 'Pamela Jackman'	*C. macropetala* 'Maidwell Hall'
	C. macropetala 'Markham's Pink'

In April there starts the season of the atragenes. These are clematis that always feature petal-like staminodes in the centre of the flower, not because of an aberration or mutation selected by man, but because this is their nature. Some botanists would hive this group off into a separate genus, *Atragene*. I am glad that most of them have so far refrained. However, it is easily recognizable as a natural group and it comprises the species *C. alpina* and *C. macropetala* from Europe and Asia; *C. verticillaris* (*C. occidentalis* var. *occidentalis*) from North America. The last can be quickly dismissed; it sometimes makes its appearance in a nurseryman's list but never seems to establish well in this country. My own plant of it was unthrifty, while it lived.

In this group the flowers nod; they hang over the plant like crowds of fairy lanterns – rather a hackneyed simile but so apt that I cannot think of any improvement. A few blooms are commonly produced at the ends of the current season's shoots in summer and autumn, but the vast majority are borne, singly, in the axils of the previous season's vines, in April and May. I first saw *C. alpina* growing wild in the Engadine in Switzerland, where it scrambled over rocks in open wood-land. That was in July, but then spring arrives late, above 2,000 metres.

The typical colouring is blue. As you would expect, a number of improved selections have been made. In recent years the number of named alpinas has vastly increased. So much so that the RHS has recently called for trials to be carried out on those currently available. I hope that this goes ahead and soon – the group is getting into a tangle as it is. That with the longest sepals is called 'Frances Rivis' (though it first, mistakenly, received its AM in 1965 as 'Blue Giant') and was cultivated by the eponymous lady in her garden at Rosehill, Saxmund-ham, Suffolk. But the original seedling was raised by Sir Cedric Morris, increased by cuttings by Hilda Davenport-Jones, and subsequently introduced and named by her. The sepals are up to 5cm long.

'Columbine' is a delightful light blue of Ernest Markham's raising, and 'Ruby' I take to be the plant he exhibited at the same time in 1937 as *C. macropetala* 'Ruby', though it is clearly an *alpina*. The colouring is a dusky reddish-mauve, not entirely satisfactory. 'Pamela Jackman', named after Rowland Jackman's daughter, is a fine deep blue. I have it clambering over a pieris, whose shrimp-red young foliage is at its brightest just when the clematis is flowering. They contrast well. Another which can be worked into a colour scheme is 'Burford White', a good clear white which like *C. alpina sibirica* has pale foliage. It was raised by one of Treasure's customers, but we have no further details.

C. alpina sibirica has white or off-white flowers and its foliage is a paler green. A correspondent in Norway tells me that a small colony of it grows there, near Lillehammer, in quite severe conditions (minus 30 degrees C), that it is not found at all in Sweden but recurs in Finland, again in a very restricted locality. Of course it has a wide Asiatic distribution. The cultivar to grow, of this, is 'White Moth'. I think it is really my favourite of the whole atragene group. In *C. alpina* itself, the staminodes are short and scarcely show, but 'White Moth' is more like *C. macropetala* in having long, conspicuous staminodes that fill the bloom out in a plump and pleasing fashion. Its white flowers go very appropriately with its pale green foliage, and it flowers in mid and late May after the main alpina season is past. It is less vigorous than its team-mates but no less healthy.

C. macropetala was introduced from China by Purdom and later by Farrer (around 1914), and received the AM on 24 April 1923. It is more showy than *C. alpina* because of its full flower and the extruded staminodes. Its colouring can be indifferent, but most nurseries trade in good colour forms only, whether they call them by special cv names like 'Lagoon' or 'Maidwell Hall' or not. I got *C. macropetala* without frills from one source and 'Maidwell Hall' from another and they seemed to me identical.

C. macropetala 'Markham's Pink' (sometimes incorrectly listed as 'Markhamii') was of Ernest Markham's raising and gained an Award of Merit on 5 March 1935 – probably a forced plant. It is rosy-mauve and purple (not pink, of course), and contrasts well with a blue form of *C. macropetala*. Vita Sackville-West had the idea of growing the two together in an Ali Baba jar, from which they were allowed to cascade at their will. The idea was a good one and has been much repeated.

As a group, then, these atragenes are remarkable for their gaiety and freshness. They will thrive as well in a sunless position as elsewhere, and they are excellent mixers with shrubs of moderate vigour to match their own. If they fail to flower it is probably because the birds have pecked out and eaten their developing buds, though late frost on an exposed plant could also be lethal. The caterpillars of one of the winter moths can also do considerable damage, by night, in late winter and early spring to both these and the montana's, which is the next group on my list.

THE MONTANA GROUP

C. chrysocoma

C. chrysocoma sericea

C. montana

C. montana 'Grandiflora'

C. montana rubens

C. montana wilsonii

C. montana rubens

    'Elizabeth', 'Picton's Variety', 'Pink

    Perfection', 'Tetrarose', etc.

The montana group of clematis is easily recognized. It stands apart from the rest in appearance, and its members are all so vigorous that they may be set the popular task of climbing into and draping quite large trees. If you have a number of trees you want pepped up in this way, you will soon discover that the variations on the montana theme are all pretty much alike. All are either pinky-mauve or white, all flower in May or only slightly later, all have four sepals as a general rule. At flowering they are exceedingly prolific but the display lasts little longer than a fortnight. I don't hold this against them. After all, you can grow them in places where the alternative would most probably be to grow nothing, so everything they give can be regarded as a bonus.

Like the atragenes, the group has been inundated recently with new named clones and these too will need sorting out some time. Let us start with *C. montana* itself and its natural 'pink' variety, *C. montana rubens*. The former was introduced by Lady Amherst from the Himalaya in 1831, the latter by E. H. Wilson from China in 1900. Bean optimistically describes the latter as 'a beautiful rosy red'. It is rosy-mauve at best, but often of so washed out a colouring that you wish it were white. The reason for these poor forms (I have several in my garden) is that *C. montana* seeds itself so readily and one is loath to get rid of a self-grown seedling if it has placed itself well. There are many named clones and these will predictably give more satisfaction than the self-sown product, but you cannot rely on these clones being true to name, even so. 'Pink Perfection', as I once had it, was indistinguishable from 'Elizabeth', but Fisk makes clear distinctions between the two.

Clematis montana is white. You don't have to call it montana alba or anything of that sort: it is white by definition. The gardening public is often unaware of this and think that, if they order a *C. montana*, it will be pink, which is, of course (but why, all the same?) the more popular colour. If they want a pink montana they must say so.

Another misconception arises over the question of scent. Because the pale pink clone of *C. montana rubens* called 'Elizabeth' is sold mainly on its strong vanilla scent, there is a widespread notion that *C. montana* is not generally scented. But, on the contrary, it often is, both in the white species and in many clones of *C. montana rubens*. I have a far more strongly coloured *C. montana rubens* than 'Elizabeth' but its scent is terrific; so is that of the old white *C. montana* in my rose garden, and its blooms are 7 or 8 cm across – but the sepals are narrow.

This is another point to watch out for: width of sepal. I don't mind them fairly narrow, myself: they look like the sails of a windmill, but you may prefer a less lean and hungry flower, something more prosperous and full. *C. montana* 'Grandiflora' is that; so are the pink clones 'Elizabeth' and 'Tetrarose'; also *C. chrysocoma* and *C. c. sericea* and *C. × vedrariensis*.

Despite its pale colouring, I do think 'Elizabeth' is a good clone with a very sizeable flower. Remember, though, that if you plant it in a shady position (as I have, with an enormous mulberry tree on one side of it and an enormous apple on the other) the flowers will come practically white. You may not mind this and I do not, but these words may prevent you from telling the nurseryman that he supplied you with the wrong variety.

'Picton's Variety' is as deep a colour as I have seen (but not so deep if grown in shade, funnily enough – you might think that absence of bleaching sunlight would give rise to stronger colouring, but this is not so) and I have a high regard for this cultivar, which often has five or even six sepals in its flower instead of the normal four. It is not quite as vigorous (though just as healthy) as the general run of *C. montana rubens*, and this can be an advantage if you want a pink montana but don't want to be smothered by it. Not much scent, here. Incidentally the pigmentation of these pink montanas shows very clearly in their young stems, leaf stalks and foliage: the deeper the flower colouring, the deeper the dye throughout the plant.

A popular clone at the moment is 'Tetrarose'. It was produced artificially in Holland, the plant's chromosome number being doubled by treatment with colchicine. Its flowers and foliage are hence larger and coarser than have heretofore been seen. Unfortunately it has been propagated from seed and therefore several forms exist. I have also noticed a marked tendency to die-back in young shoots during the winter. You can have it (probably you already have and I shall receive a stream of angry letters).

Two more points about *C. montana* and *C. montana rubens* are worth noting. There are some clones that, from young rooted cuttings, take several years of growing and settling down before they start flowering. In general, montanas flower abundantly from the first year and never look back, so the other condition can be a worry. The perplexed owner wonders if it needs pruning. On the contrary, he should leave well alone. In due course, his plant will flower as freely as any other and once it has started it won't have second thoughts. This was the case with the plant of *C. montana* 'Grandiflora' that I used to own – but I'm sure there are many clones around under this name and that some of them are precocious flowerers. It was also the predicament of a pink montana that I bought from the late Hilda Davenport-Jones – a great plantswoman, and I did not regret my purchase but it was a slow starter.

It grew into a 'Blenheim Orange' apple tree. This was not a clever association and brings me to my second point. Montanas flower in May at much the same season as apples. If you fling a pink montana into a pink apple tree and there is pink apple blossom all around, no one will look at it once, let alone twice.

A number of new montana clones have appeared recently, many of which have been introduced by Jim Fisk. We are not familiar with them all. 'Alexander' is a creamy-white and quite sweetly scented. It is a vigorous plant with pale foliage although slightly deeper than *C. montana* itself. It takes time to settle down to flowering. 'Marjorie' is described by Fisk's as semi-double with creamy-pink outer sepals and salmon-pink inner ones. Neither of us has seen a large, established plant, but smaller four-year-old ones with Tom have not yet flowered and other growers have suggested that it is not particularly free. 'Freda' has proved to be a good deep colour which rivals 'Picton's Variety', and the young plants that Tom has seen (it was introduced only in 1985) have flowered well. Another point in its favour is its attractive bronzed foliage. 'Vera' is a particularly vigorous plant, even for a montana. The foliage is bronzy but rather coarse. It has pink flowers and a reasonable scent.

Now we come to *C. montana wilsonii*, which is another natural variety introduced by the eponymous Wilson. It is immensely popular as a follow-on to *C. montana* itself because, as you will read in every textbook except mine, its flowers, to quote Bean, 'appear in July and August'. He also states that it has larger white flowers, 3ins in diameter. The nursery firm of J. Veitch & Sons called it 'the autumn-flowering *C. montana*'!! In *Curtis's Botanical Magazine* (t.8365, 1911) its flowering season is given as two months later than *C. montana*, and it is said to flower from the nodes of the previous season's wood. How it could do this as late as July, when the whole plant would be a mass of young growth, is hard to imagine.

There are at least two completely different clones purporting to be this clematis going around. There is a breathtaking example of the one I feel confident to be the genuine article growing on Keillour Castle (home of my late cousin, Mary Knox Finlay) in Perthshire. I have seen this plant on many occasions in early summer and it normally starts flowering in mid-June. It is slow to get going, and if you see it at the start of its flowering period you will think what a miserable thing: greenish white and spidery. But the whole plant, from roof to ground – a matter of 12m (40ft) is one sheet of white and giving off great wafts of hot chocolate aroma. It is most impressive. The flowers are actually 9cm in diameter but the sepals look narrow because, in the mature flower, their margins roll back. The stamens make a prominent brush. Now this is habitually at its peak around 24 June, which is certainly a month later than most montanas. Not the two or more months some authorities would lead us to expect, but still very well worth cultivating as a sequel to *C. montana* and without overlap.

The plant I originally bought from Jackman's may well have been this, but I was so disappointed in the miserable, greenish-white flowers of the young plant that I discarded it (and so did Jackman's) before I had given it a real chance and allowed it to develop to maturity.

Questions of identity still plague us when we turn to *C. chrysocoma*, which was discovered in Yunnan, W. China, by Delavay in 1884 and presented to Kew by Monsieur de Vilmorin in 1910. It cannot have been long there when Bean wrote of it (*Trees and Shrubs Hardy in the British Isles*, 1st edition, 1914) as a 'semi-woody shrub 6 to 8ft, perhaps more, high'; he also writes of its 'short, erect habit' and of it being 'rather tender at Kew'. These entries remain unaltered in the latest, revised edition of Bean of 1970. They bear no relevance to the plant as we know and grow it, which climbs with a will to 6 or 9 metres and appears – at any rate in maturity – (I have had winter casualties among young plants) to be as hardy as *C. montana*. It figured in *Curtis's Bot. Mag.* (t. 8395) in 1911 and was said to have died to the ground at Kew in its first winter (1910–11). Its illustration has an unreal and stylized appearance but gives a distinctly chunky impression. I think the plant was written up before its appearance and capabilities were properly known. Seed of the shrubby *C. chrysocoma*, found in the wild in China, has recently been introduced into the UK and it will be interesting to see what form the mature plants take.

C. chrysocoma is a very beautiful clematis with more quality (that indefinable something) than *C. montana*. Young leaves, stems and flower buds are all covered with soft down (chrysocoma means with golden hair). The leaves are very handsome, with their bold lobing. The flowers are borne on extra long stalks and they hence present themselves importantly. They are a soft but definite rosy mauve, of an excellent shape with four broad sepals, but no scent.

It is a habit in this species, especially when growing actively in youth or because it has been pruned, to carry a second, spasmodic crop of extra-fine blossoms on its young trails. This explains how it managed to get an Award of Merit on 7 July (when exhibited by Ernest Markham in 1936).

C. spooneri was recognized by Rehder and Wilson, who described it under this name, as being a close relation of *C. chrysocoma*. It is now generally accepted (i.e. Lloyd now accepts it) as being a variety of the latter, and should be known as *C. chrysocoma* var. *sericea*. This seems sensible. The leaves are similarly broad for their length, downy on both surfaces. The flowers are bold, 8cm across, odourless. In the clone I acquired from Captain Collingwood Ingram there is a slight pinkish flush along the central veins of each sepal, which seems appropriate to the species, but, as mentioned above, the clone offered by Hillier *et al.* is pure white. There is no scent in either case but plenty of vigour.

The cross between *C. chrysocoma* and *C. montana rubens* is called *C.* × *vedrariensis*. In trade catalogues you often meet it as 'Spooneri Rosea'. The original cross was made by Messrs Vilmorin, the famous Versailles nursery firm, and they first exhibited it in 1914. Similar crosses have been made on a number of occasions since, sometimes using *C. chrysocoma sericea* as one parent with *C. montana rubens* as the other. Various clones such as 'Highdown' and 'Hidcote' have been named, though the latter appears to be a good form of *C. montana rubens* pure and simple.

C. × vedrariensis received an AM when exhibited by Ernest Markham on 9 June 1936. It is always an excellent clematis – just about the best going for size and quantity of blossoms of a good pink colour on an ultra-vigorous plant. I find it roots and gets going from cuttings much more slowly than the montanas themselves.

THE TEXANS

C. addisonii	'Duchess of Albany'
C. crispa	'Etoile Rose'
C. pitcheri	'Gravetye Beauty'
C. texensis	'Sir Trevor Lawrence'
	C. viorna

Clematis texensis is potentially one of the most exciting species within the genus. Its colouring (and, incidentally, the size of its flowers) varies greatly and may be pinkish or purplish, but at its best is pure, bright red. The flower form is intriguing, too, so it is no wonder that hybridists both here and on the Continent were busying themselves at the turn of the century on breeding texensis features into the modern garden hybrids. Their efforts met with considerable success but came to an abrupt halt in the early 1900s. The clematis was suffering a temporary eclipse at this time, mainly on account of wilt disease and its serious effect on the clematis market which, in its turn, put a damper on developments in clematis breeding from which the plant has never fully recovered. We know more about wilt disease and how to live with it, at last, and it is to be hoped that the charm and appeal of cultivars like 'Etoile Rose' and 'Duchess of Albany' will encourage a few enterprising clematis fans to probe further in the texensis field. There is obviously scope for development, here.

Clematis viorna is usually taken as the type for a group of clematis from the South Eastern United States having urn- or pitcher-shaped flowers consisting of four thick, leathery sepals with connate (i.e. joined) margins, opening out only at the tips. This section is sometimes called *Viornae*, sometimes *Urnigerae*. The plants are climbers but often herbaceous or sub-shrubby; that is, dying back a good deal if not right to the base, in winter.

Of *C. viorna* Bean writes, 'Although interesting and curious, this species is not particularly attractive.' If our interest and curiosity are aroused, are we not attracted? However, Bean is clearly using the word attractive in its popular sense and one must admit that small flowers of subfusc colouring are not calculated to make the onlooker put on his dark glasses, gasp, jump back in horror or upwards for joy. I'm fond of this species and have grown it for many years, but the conspicuous seed heads are its main feature. I have likened them to tropical spiders (though they have far more than the quota of eight legs to which even the largest and most tropical spider must, presumably, be limited).

In another species of this group that I have seen, *C. crispa* to wit, the colouring was pale bluish, rather like *C. heracleifolia davidiana*, and the sepals more spreading than usual among *Viornae*, and twisted. I have also clapped eyes on *C. addisonii*, which is rosy-purple outside, cream within and along the outer margins. Most of these clematis come readily from seed (*C. viorna* certainly does), and one can sometimes locate this in the lists sent out by Botanic Gardens.

C. pitcheri was supposed to have been used as well as *C. texensis* in the late nineteenth-century hybridizing work done on the Continent, but there was some confusion of identity here, and it seems probable that *C. texensis* was, in fact, being used all the time.

Let us get back to this species and see what it has done for us and we for it. It is not supposed to be fully hardy, but seems to be pretty nearly so in Britain and is, as I have seen it at Malahide Castle, beautiful as well as curious and interesting. But it rarely comes on the market, and when it does you are usually offered a seedling which is more than likely to give disappointing results. The fact is that *C. texensis* is a bit of a pig to propagate from cuttings. The little seed of it that has come my way has failed to germinate.

C. texensis is said to have been introduced into Europe in 1868. It was taken up and improved by Max Leichtlin (I owe the ferreting out of this information to Hugh Thompson), who had his own private botanic garden at Baden-Baden. In those days the species was known as *C. coccinea*. In William Robinson's publication *The Garden*, 8 August 1903, Leichtlin wrote the following note: 'Of *Clematis coccinea* I have by constant selection and constant sowings now produced a fine seedling, the flowers of which are three times the size of the wild plant.' This would seem to be a different plant from the *C. coccinea major* referred to by Bean, of which a plant once grew at Abbotswood. The flowers of this were only half as large again as in the normal species. Both the one and the other have disappeared, in any case.

But Max Leichtlin was also the first person, around 1885, to cross *C. coccinea* with large-flowered hybrids. He passed on the handsome deep reddish-purple variety so obtained to Louis Späth, after which we hear no more of it. (This cross is recorded in *Les Clématites* by Boucher and Mottet, 1898.) Leichtlin was also responsible for the introduction of *C. coccinea* to England in 1880.

In his notes for the revision of Moore and Jackman, A. G. Jackman pencilled in: 'In the summer of 1890 another interesting new type was raised at the Woking Nurseries by crossing 'Star of India' with pollen from the American species *C. coccinea*, which resulted in the introduction of the pretty campanulate hybrids, 'Countess of Onslow', 'Duchess of Albany', 'Duchess of York', 'Grace Darling', 'Sir Trevor Lawrence' and 'Admiration', showing great variety in shades of colours not previously obtained in Clematis, intermediate in sizes between the two parents, of more open campanulate form than *C. coccinea*, but partaking very closely to it in the substance of the sepals and foliage and also in the character of the growth, being of a sub-shrubby habit.'

These clematis became known as the Wokingensis hybrids. Only 'Duchess of Albany' and 'Sir Trevor Lawrence' remain certainly with us. A variety is marketed as 'Grace Darling' in New Zealand, but the coloured transparencies I have seen of this exactly resemble 'Duchess of Albany'. A. G. Jackman describes 'Duchess of Albany' as we know it today: 'bright pink, deeper down the centre, softening down to a lilac pink round the margin...'. 'Grace Darling' he describes without differentiation as 'bright rosy carmine'. 'Countess of Onslow' is a puzzle. I once owned what was supposed to be this variety (I had it from Jackman's). It had bell-shaped flowers with four sepals, pale pink at the margins deepening to cerise at the centre on both surfaces, but the colour transition was more marked on the outside. A. G. Jackman, however, describes the flower as 'bright violet purple with a broad band of scarlet down the centre of each sepal'. A very different matter. Anyway, I lost my plant of whatever it was in the 63 winter. It rather looks as though we have lost both the Countess and the pseudo-

Countess. She still causes many a wild goose chase. We have heard of several occasions when enthusiasts have followed up sightings of likely sounding texensis cultivars in old gardens. They usually end up finding 'Duchess of Albany'. A recent claim suggests that the plant may still exist in New Zealand, but I am told that a plant import licence into the UK now costs £180!

Of the two Wokingensis hybrids left, 'Duchess of Albany' is a highly satisfactory clematis. Its bells are rather elongated and it is as near pink as you will find in a clematis. It dies back but not all the way back in the winter, and can make 3 or 4 metres of growth in a season. Its bells are held upright (like the blooms of a lily-flowered tulip) in an open situation and it associates particularly prettily, as I have seen it in John Treasure's garden, with the powder-blue panicles of *Ceanothus* 'Gloire de Versailles'.

I was responsible for rescuing 'Sir Trevor Lawrence' from oblivion. It grows at Sissinghurst Place, Kent, in the garden belonging till recently to Mrs Lindsay Drummond, was planted by her mother-in-law years ago and was fortunately labelled with one of those embossed metal efforts that usually outlast the plants they are identifying. Mrs Drummond allowed me to take cuttings, which root easily. The plant grows for me over a *Cotoneaster horizontalis* and is a real eye-catcher; shaped like 'Duchess of Albany' and with four, five or six sepals but of a luminous cherry-red colouring. A. G. Jackman's 'bright crimson' will do. Sir Trevor Lawrence, the man, was for long President of the RHS.

Now we come to what is undoubtedly the pick of the bunch, though always in short supply (for no valid reason that I can see). Unlike the last two, the flowers of 'Etoile Rose' are nodding; they are open-bell shaped and rather short. The plant dies right to the ground or within 0·4 metres of it each winter, but is immensely vigorous (3 or 4m in a season), starts flowering early or mid-June and continues for two or three months with remarkable freedom and reliability. Like other texensis hybrids it is subject to mutilation by mildew, so I would recommend an open site unless you are prepared to take protective measures. The flower has four sepals which are silvery pink at the margins, deep cherry purple in the centre.

Nearly all the plants now being grown originated at Abbotswood, Stow-on-the-Wold, where it was introduced by the then owner, Mr Mark Fenwick, in 1929 or 30, from a Monsieur Chenault in France.

Fenwick showed it to the RHS on 4 July 1939 as *C. viticella* 'Etoile Rose', and it was sent for trial to Wisley. War intervened. The Sunningdale Nurseries, through Mr Graham Thomas, subsequently obtained stock from the Abbotswood plant and got an Award of Merit for it on 7 July 1959. Hugh Thompson traced the early history of this cultivar (RHS *Journal*, No. 95, pp. 189–90). It was described in the List No. 158 (1903) of V. Lemoine et Fils, and Lemoine was the raiser. He gave the parentage as *C. × globulosa* and a variety (unspecified) of *C. viticella*. The parentage of *C. × globulosa*, also raised by Lemoine, was *C. douglasii* var. *scottii* and *C. texensis*. If you consider the open bells and the way they are presented, it is not so surprising to find that *C. viticella* played a part in the making of 'Etoile Rose'. *C. douglasii* var. *scottii*, the purely herbaceous, outlying species from the west side of the United States, is a surprise, however.

Finally, 'Gravetye Beauty'. We know that nearly all the clematis raised and exhibited from Gravetye by William Robinson and Ernest Markham came to them originally as seedlings from Morel, and this one was no exception, as was expressly stated when exhibited to the RHS (it received an AM) by Ernest Markham in September 1935. It was raised by Morel of Lyon and introduced to England by William Robinson in 1914. Morel had been using *C. texensis* (*C. coccinea*) for hybridization but what the other parent was is hard to guess. The blooms start the typical elongated-bell shape of the Wokingensis hybrids, but then open wide into a narrow-limbed star of four, five or six sepals with inrolled margins. The shape at this stage is not particularly good, but all in all this is an excellent clematis, of a rich crimson colouring. Often it dies back to the ground in winter, but when it keeps any old wood it starts flowering in June; otherwise not till August. It is a thoroughly good doer but does not make an inconvenient length of growth and is hence suitable for association with shrubs of moderate vigour such as *Paeonia delavayi*.

In the *Revue Horticole* of 1904 (p. 308) there is a delightful plate of three of Morel's *texensis* hybrids, the three other parents used in these crosses being 'Gipsy Queen', 'Comtesse de Bouchaud' and 'Ville de Lyon'. Unfortunately these hybrids have disappeared.

In a letter to me from John Treasure written in 1971, he remarks that texensis hybrids seldom set seed on their own but that when he pollinated his 'Etoile Rose' with *C. viticella*, the three blooms thus treated set good seed. This has led to the production of *C.* 'Pagoda' (see p. 190).

────────── THE VITICELLAS ──────────

C. campaniflora	'Little Nell'
C. viticella	'Mary Rose'
'Abundance'	'Minuet'
'Alba Luxurians'	'Pourpre Mat'
'Elvan'	'Purpurea Plena'
'Etoile Violette'	'Purpurea Plena Elegans'
'Kermesina' (syn. 'Rubra')	'Royal Velours'
	'Venosa Violacea'

And so we move easily forward to *C. viticella* and its satellites. *C. viticella* has a place in the parentage of so many hybrids that I am reluctant to use its title for a group that is bound to be arbitrary. Why is 'Huldine' not here, you may ask, why not 'Margot Koster'? If they are being excluded, why not 'Etoile Violette' also? My only answer is that I have grouped them as I have for my own convenience. Full Stop.

John Gerard wrote of *C. viticella* 'it climbeth aloft and taketh hold with his crooked claspers upon everything that standeth nere unto it'. As you would expect of a species that has been cultivated in this country both for itself and as a stock for grafted varieties, since the sixteenth century; that is, moreover, easily raised from seed and that grows wild almost on our doorstep in South Europe, *C. viticella* is very variable even before you start crossing it with this and with that. It has far greater elegance in the manner its flowers are presented than any of the cultivars we shall consider. Flower size varies a lot: in a good strain flowers may be 8 or 9cm across, but they look less, for there are only four sepals and they are narrow or look narrow, their margins being often rolled back. The beauty of this species, to my way of thinking, is that the flowers all nod and that they are held well out from the plant on extra-long stalks. Stalks and flowers are purple, the colouring being more intense on the upper side of the bloom, so it is a pleasure to look down on it.

In none of the viticella hybrids (and I feel sure they are all hybrids rather than selections or mutations) is the same elegance retained; none have the long stalk, and the flowers, if they nod at all, do not do so consistently. They look out at you boldly rather than down, demurely.

Perhaps 'Alba Luxurians' comes nearest to its parent in charm and elegance. We have already remarked on the narrow boundary between leaves and sepals in *Clematis*. The earliest blooms of 'Alba Luxurians' are entirely green, some of their segments broadened and lobed and closely resembling leaflets. The normal condition in the main crop of blossom is a white sepal with green tips and dark anthers. Sometimes the flower is white throughout. There is a tendency in the foliar flowers for the sepals not to be shed, when dying, but to remain in a brown condition on the plant. This doesn't matter too much. It received the Award of Garden Merit in 1931.

Bean opines that this clematis was probably raised at Veitch's Coombe Wood nursery. Its name was pencilled into Moore and Jackman by A. G. Jackman for his proposed revision, so it has been in existence for at least seventy years. There were whitish cultivars of *C. viticella* style around before this.

From the time it was first brought into cultivation, *C. viticella* has been curiously ready to produce double sports (as one imagines the early examples must have been) and hybrids. Gerard cultivated both *C. viticella* and the double-flowered 'Plena': 'they grow in my garden at Holborn and flourish exceedingly'. (Why couldn't he have made some less banal observations?) Bean says of the double-flowered variety that it is one of the least attractive 'owing to an excessive multiplication of the sepals, which gives the flower a heavy, lumpy aspect'. He was wrong in writing of only one double variety and one wonders which of the many that were current in his day he was thus anathematizing. Graham Thomas is as fastidious a gardener as we have among us today, and writes of the one he called *C. v. elegans plena*: 'As a rule, double clematises do not appeal to me but this has such a neat rosette of flattened petals, about $2\frac{1}{2}$ inches across, and its colour is of such a soft lilac purple, reminiscent of some of the old roses, that it seems too good to be lost' (RHS *Journal* No. 90, 1965, p. 402). The lilac-purple, old-rose flowers that he describes are, indeed, almost brown, and very intriguing at close range. The plant is vigorous and gives no trouble in cultivation. It grew in Mark Fenwick's garden at Abbotswood and, like 'Etoile Rose', was put into circulation again by the Sunningdale Nurseries through Mr Thomas's connection with that firm. They listed it as *C. viticella elegans plena* but Treasure's of Tenbury have subsequently re-named it 'Purpurea Plena Elegans'. This needs further explanation and a new paragraph.

At the turn of the last century and this, Lemoine of Nancy and Morel of Lyon were bringing out a number of double viticellas. Two of the latter were described by Edouard André in the *Revue Horticole* of 1899. These were *C. viticella* 'Purpurea Plena' and *C. viticella* 'Purpurea Plena Elegans'. Treasure's have decided on the latter as the possible/probable original of the Abbotswood 'Elegans Plena'. The other name, 'Purpurea Plena', they gave to a double viticella of which they received blooms and propagating material from a garden in Dorset (or Devon; they forget which). I never saw this and they subsequently lost their stock of it, but I am given to understand that its flowers are (or were) of a bluer shade of purple than the other. André doesn't help much, because in describing the flower colour of 'Purpurea Plena' and 'Purpurea Plena Elegans', he in both cases gives it as 'violet-pourpre foncé' – deep violet-purple. All in all I think it may fairly be said that you pay your money and you make your choice. We could be growing Morel's clematis or we could be growing others – of his or of someone else's. Peveril Nursery have recently introduced a double viticella which they have called *C. viticella* 'Mary Rose'. They found it growing by a manor house in Devon and believe it to be of great age. It differs from *C. viticella* 'Purpurea Plena Elegans' in being altogether bluer in colour (they describe it as 'smoky amethyst') and having a smaller, more spiky, flower. An interesting and attractive clematis, we have so far failed in our attempts to discover more from Peveril's about its history or whether it is implicated in the 'Purpurea Plena'/'Purpurea Plena Elegans' mystery.

At this point I shall mention *C. viticella* 'Elvan', which is another from Peveril's. The flower comes closest to *C. viticella* and is carried on long stems in the nodding manner which is so attractive in the species and none of the others in the group quite manage. Vigorous and prolific, the colour is quite close too: purple, but with a blaze down each sepal. A good introduction, we could do with some more. There are other viticellas around but they are not widely known. 'Blue Belle' has recently arrived from the Continent but we have yet to see it in flower.

A similar muddle with 'Kermesina' and 'Viticella Rubra' has been resolved and 'Kermesina' is now accepted as the correct name. As with some other named clones, seedlings may still confuse the issue. At Sissinghurst Castle they have several red viticella hybrids that are similar but not quite the same. They are unnamed and no one can say just what they are. The original 'Kermesina' was raised by Lemoine.

This clematis is first class: of a bright colouring, effective and prolific. Also very vigorous, easily making 4m of growth in a season. 'Abundance' comes nearest to this in colouring and is very similar in size and habit. The colour is a bluer shade of red and lighter. G. S. Thomas calls it 'warm claret-crimson', but I should be deeply suspicious of any claret with so much blue in it. Generally speaking this is a most satisfying clematis, but I must in honesty record that with me it flowers shyly, although its vigour is terrific. I have it in heavy soil – which it obviously likes – and it does not receive much sunlight until after noon. If it can behave like this in one garden it may in another.

'Abundance' was one of the seedlings given by Ernest Markham to Jackman's and was named by them. There is therefore a strong probability that it was one of Morel's raising and was among those obtained by William Robinson before the First World War at the time when Morel turned from nursery work to garden design. The same provenance is likely to apply to 'Minuet' and 'Little Nell', both named and introduced by Markham, whose wife's name was Nell.

Neither of these varieties has received an award and I can quite understand this as they are essentially garden plants, effective in a garden setting but not when divorced from it. Both have typical small viticella flowers and are pale in colouring – basically white, in fact, but 'Minuet' is margined in light purple and 'Little Nell' in a paler mauve. You may suspect that they will look pasty and anaemic but this is not actually the case.

'Royal Velours' is the darkest of the dark. It has a similar history to the above with a strong suspicion of Morel as raiser. William Robinson exhibited it to the RHS in August 1934, but whereas he got an award for 'Huldine' on that day, 'Royal Velours' was passed over. But it got the AM when shown by Walter Bentley of Quarry Wood on 20 July 1948. There is a famous specimen at Hidcote. Graham Thomas wrote

(RHS *Journal*, 1965) of its being of intense dark wine-purple, velvety texture and of annually causing much comment in association with the orange-red fruits of the Japanese wine-berry (*Rubus phoenicolaseus*). Very dark clematis do need careful placing to be effective.

'Pourpre Mat' is probably another such, though I have not seen it and it is no longer offered. This was definitely one of Robinson's introductions from France (Morel) as was stated when it received an AM on 10 September 1935. Mr Rowland Jackman told me, however, that it is disappointing and that AMs were more freely granted in those days! Its blooms were 5ins across, so it was one of the larger hybrids ascribed to *C. viticella*.

So are 'Etoile Violette' and 'Venosa Violacea'. 'Etoile Violette' is of much the same purple colouring as *C. × jackmanii*, but, although it can have only four or five sepals, most of the time it has six and the flower is of a far more pleasing outline. Furthermore, its eye of creamy stamens gives a focus to every bloom. 'Etoile Violette' has an earlier season than many clematis flowering on their young wood: it is commonly at its best in the first half of July. It was raised and introduced by Morel around 1885.

'Venosa Violacea' (which is distinct from the earlier 'Viticella Venosa') was raised by Lemoine and might be compared with an enlarged 'Minuet'. The ground is white overlaid with reddish-purple veins in the centre, but uniformly purple at the margins. The flower is quite differently shaped from 'Minuet', however, with its five or six boat-shaped sepals. An intriguing clematis.

Finally we can return to an untrammelled, unimproved species, *C. campaniflora*. It comes from Portugal and is very close to *C. viticella* but the flowers are almost white with a faint suggestion of violet. Rowland Jackman well described it as 'a dear little flower, like little nodding, delicate mauve bells'. They are, in fact, quite open bells, only 2 or 3cm across. A vigorous plant, John Treasure grows it into a large conifer close to the front of Burford House. It is perfect for the position and looks splendid, its dainty flower quite belying the vigour of the plant. William Robinson, in *The Virgin's Bower* (p. 15), also observed that it was a particularly elegant climber. He saw a blue form in a garden at Weybridge but failed to procure it. So have we all.

————— LATE-FLOWERING SPECIES —————

C. brachiata	*C. rehderiana*
C. fargesii var. *souliei*	*C. serratifolia*
C. flammula	*C. tangutica*
C. maximowicziana	*C.* × *triternata* 'Rubro-marginata'
C. orientalis	*C. vitalba*

My next group is homogeneous only in that it consists almost entirely of species and that they are all late-flowering, on their young wood. So, although mainly rampant, they can be shortened back as hard as you please each winter. On the other hand, if you don't prune them they will get along quite nicely as they would in nature. Untidily, maybe, but over a small tree or large bush they will not appear untidy, except in the dormant season. And they will start flowering a good deal earlier.

If you want a yellow clematis, this is where you should seek. The brightest, lemon (or even buttercup) colouring is found in *C. orientalis* and *C. tangutica*. Since the latter is often considered as a mere variety of the former (and I find myself, in many instances, hard put to it to say which is which), I will take *C. orientalis* first (although recently a classification change has been mooted, see p. 189). It was originally introduced in 1731 and its synonym, *C. graveolens*, is often met with in literature. *C. orientalis* has such a wide distribution, from the Caucasus to N. China and Manchuria, that its variability is not to be wondered at. In modern times, gardeners took little notice of it (preferring the forms referred to *C. tangutica*) until Ludlow and Sherriff made their remarkable introduction, under the number L & S 13342, from their Himalayan expedition of 1947.

C. orientalis L & S 13342 has delicate, finely cut foliage that makes a misty fretwork of green, and globular flower buds that open into four thick sepals having the colouring and thickness (or an impression of the thickness) of lemon peel. Unfortunately the name 'Orange Peel' is sometimes applied to this clematis or to seedlings from it, but the colouring is only ever orange on fading blooms in a sunny position. It is not typical.

I find it difficult to advise on whether you should prefer this clematis (it received the AM when shown by Messrs Ingwersen on 10 October 1950) to a good form of *C. tangutica* or not. The foliage is certainly superior. The plant itself has similar vigour but its hardiness record is not unstained. I have known a number of cases from the north where plants have died in the winter, though whether from cold or from another cause I cannot say. Although it may flower excellently in a sunless position in the south, it is very apt to run to leaf under these circumstances. If there is any tendency in this direction, let up on the pruning. Indeed, light pruning gives a very long season, from June onwards. I may say that on a sunny wall in north-west Scotland (Dundonnell House, in Wester Ross) *C. orientalis* L & S 13342 gives a splendid account of itself.

It should always be raised from cuttings, which root easily and quickly but also rot quickly, having rooted, unless potted and hardened off promptly. It has, however, often been propagated from seed, which gives a variable and generally inferior product. If you buy *Clematis* 'Orange Peel' or the Orange Peel Clematis, you have no guarantee of its pedigree. If you buy L & S 13342 you should, from a conscientious nursery, be safe.

Treasure's market the progeny of a self-sown clematis in John Treasure's garden that is probably a hybrid between *C. orientalis* and *C. tangutica*, calling it *C. orientalis* 'Burford Variety'. This is showy and vigorous. John Treasure doesn't like it. As the price of flower power one must always expect a certain coarseness in such hybrids. Another hybrid of the same parentage (*C. orientalis* L & S 13342 was the seed parent) has been named 'Bill Mackenzie' after the retired Keeper of the Chelsea Physic Garden. It originated at the Waterperry School of Horticulture in 1968 and won an AM when (cleverly) shown by Valerie Finnis to the RHS on 7 September 1976. Again it has flower power and is vigorous. Unfortunately, this too has been raised from seed from time to time and you need to be just as careful when you buy as you would with L & S 13342.

While comparing the two species, a drawback to *C. orientalis* that should be mentioned is that its seed heads are not striking in the manner of *C. tangutica*. It has been said of both species that their flowers are sweetly scented. I have tried umpteen samples without the faintest response, but the allegation cannot be safely refuted, since the closely allied *C. serratifolia* is strongly scented. Somewhere there may be scented clones of the two species under consideration, but perhaps they were so inferior in other respects as to have dropped out of cultivation.

C. tangutica first reached this country from the Imperial Gardens of St Petersburg (now Leningrad) as recently as 1898, though its home is central Asia and it has frequently been re-introduced since then. It reached France earlier. Writing in 1902 in the *Revue Horticole*, Francisque Morel says he received his first plant some fifteen years earlier. He says that (unlike *C. orientalis*) it would not flower at all if hard pruned: only off the old wood. The illustration shows an open-flowered form up to 8cm across at the mouth and very like a specimen on Jack Drake's nursery near Aviemore in Inverness-shire, from which he has distributed seedlings from time to time. The sepals are some 5cm long.

However, *C. tangutica* var. *obtusiuscula* was introduced by Wilson in 1908 and Purdom in 1910 from Szechwan and Kansu in W. China, and this has smaller more bell-like flowers. Not so showy, but displayed well on long stalks. From seed collected in 1911 and sown at Goring-by-Sea in Sussex in 1912, Frederick C. Stern won an Award of Merit in July 1913 at his plants' first flowering. All these yellow-flowered clematis are very quick and easy from seed. I have raised *C. tangutica* from seed to flower between spring and autumn in the same year.

In the report of his expedition to Kansu in 1914–15, Reginald Farrer wrote of *C. tangutica obtusiuscula* that it 'unfurls a coil almost as long as its name over the river-shingles of all the streams above Jo-ni, ascending to about 10,000ft on the fringes of the alpine coppice. In August it is all a dancing carillon of big, yellow bells like gay golden Fritillarias, succeeded in November by the most voluminous fluffs of soft silver that I know among these Clematids' (RHS *Proceedings*, 1921).

Most of the plants you see in gardens are undistinguished seedlings but good even at their worst. Lightly pruned, they start flowering in June, and the succession produced over the next four months is soon joined by the seed heads that are such a distinctive feature in this species – grey and yellow, youth and old age, together.

You can reckon on 4 or 5 metres for this clematis and *C. orientalis*, so they cannot compete with big-bully cherries like the double white gean, but can with a large shrub or small tree, like the morello cherry.

In a way, *C. serratifolia* must be reckoned a collector's piece. Though similar to the two foregoing it is rarely as showy, but it has its points. The flowers appear in August–September and would politely be described as soft yellow, less sympathetically as wan yellow. They all come in a rush together, which makes for a good display, but this season is correspondingly short, say three weeks. One of their points of distinction is the boss of purple stamens, which are conspicuous. The other is their scent, which is strongly of lemon. The feathery styles on the ripened seeds are also charming and come in swags. If you want them to last you must pick them promptly for the house, as they soon disintegrate, unlike Old Man's Beard. There are better forms around which would raise the status of the species were they to become current.

C. rehderiana is a very pale yellow, paler than primrose, but what a clematis. Time and again journalists enthuse about it and an attractive photograph points their praises, but the trade is quite unable to meet the resulting deluge of orders. Like *C. orientalis* it is quickly rooted from cuttings, but these as quickly rot if kept close for a few days too long. If hardened off and potted as soon as rooted, there are no problems.

This is a slightly more vigorous clematis than those just described, and will grow 8m into a tree. On the other hand it can be cut hard back every winter and will always flower abundantly from August to October. The pinnate leaves are conspicuously hairy; you cannot miss this feature. The flowers are numerous in their panicles and the sepals touch at their margins to form a genuine bell-shape, the sepal tips being recurved so that there are no points. And then there is that tremendous scent of cowslips. Really delicious.

Seed was first sent from W. China (Szechwan) by Père Aubert to L. Henry of Paris in 1898, and this was flowered in 1899 but wrongly identified as *C. buchananiana*. Henry wrote it up in the *Revue Horticole*, 1905, and likened the flowers' scent to orange blossom. He was a nurseryman and had perhaps never met any flower as humble as the cowslip. Wilson re-introduced the species in 1908. He and Rehder subsequently described it as *C. nutans* var. *thyrsoidea*, but in 1914 Kew raised it to specific rank as *C. rehderiana*. *C. veitchiana* is very close but I have not met it. *C. rehderiana* received the AM in 1936 and was, some years ago, granted a well-deserved Award of Garden Merit. It can set seed in this country but is a very variable species.

I have now finished with the yellow-flowered species and turn to some that are white. Of these, *C. fargesii* var. *souliei* was another of Wilson's introductions from China, in 1911. Its leaves are bold, pinnate and somewhat hairy. The flowers on a lightly pruned or unpruned specimen open from early June onwards through the summer and autumn; never many at a time, so this is no great display, but enough to be pleasing. They are sizeable, as small-flowered species go, 5 or 6cm across and, being composed of six broad, rounded sepals, they are well shaped and not at all unlike a white anemone. It grows equally well in the north. A nice clematis. *C. fargesii* itself is not in cultivation. *C. potaninii* is closely allied and may be synonymous but, according to Jack Elliott, it has larger flowers; up to twice the size in some clones.

C. vitalba is our one native species, popularly called Old Man's Beard. It will grow almost anywhere in the UK. Tom established it in his parents' garden in acid soil 800ft up in the Durham Pennines. That was over twenty years ago and it has never looked back. It does flower late, and the seed never has the chance to set, which probably accounts for its natural range not extending north of south Yorkshire. Its little off-white flowers are moderately attractive, in quantity, but the grey seed heads are a great deal better than that. The way you see them draping rough hedges and small trees like hawthorns in the wild (generally in chalk or limestone country) should give the gardener ideas. The gardener, that is, with a largish piece of land in a rural setting; certainly not the urban or suburban gardener, for *C. vitalba* is a rough, untamed creature and its habit of self-sowing can be a nuisance in an intensively cultivated garden. But if you had an old, unpruned yew tree, for instance, you could hardly do better than encourage curtains of Old Man's Beard to swing through it. Another circumstance in which I have seen this clematis looking particularly arresting was in the branches of a larch, in November, when the larch's own deciduous foliage had changed to yellow and was caught by low sunlight. Larches are large trees with greedy roots. Not many climbers could cope, but *C. vitalba* is tough and can climb to 12 metres. Moreover, its seed heads remain in condition throughout the autumn and for a large part of the winter.

Still, there's no getting away from the coarseness of the growing plant in summer, particularly of its foliage. This is where the South European *C. flammula* (one of my favourite species) is so vastly superior. The leaves are dark green but neat and gleamingly polished. The flowers are tiny cruciform little things again, but pure white and strongly scented of meadowsweet or hawthorn – a sickly kind of fragrance that yet meets you very pleasantly on the air. Even when lightly pruned,

C. flammula seldom reaches its peak until mid-August and continues throughout September – a useful season. It is not as vigorous as vitalba and can be used to enliven large shrubs like escallonias or lilacs. The seed heads are nothing to boast of, and anyway develop so late that in most seasons the seeds cannot ripen, in this country, before cold weather cuts short their development. Flammula builds up a thick trunk that becomes hollow after ten years or so, and at this age your plant is liable to die suddenly and unexpectedly. It is not, in my experience, a very long-lived species. Cuttings are rather tricky to root, but imported seed gives excellent results; so does your own after a hot summer and an open autumn. The main germination will occur two springs later, but if you are not ready to deal with seedlings in the first year or two they will hold perfectly happily in the seed pan for several years.

I have already described the effect of *C. flammula* mixed with 'Gipsy Queen'. *C. viticella* would also make an excellent purple associate of the same flowering period and pruning requirements. Flammula with *C. tangutica* would also look nice. I rather like Moore and Jackman's recommendation for siting it 'on some rocky eminence, where, being allowed to assume a decumbent habit, its myriads of pure white blossoms seem to pour down the declivities like masses of drifting snow, at the same time embalming the air with their fragrance'. Few of us have rocky eminences to play with, but obviously the author must have seen it growing in this way (perhaps in the wild) to be able to communicate such a vivid pen picture.

The plant called 'Rubro-marginata' (FCC 1863) was at first known as *C. flammula roseo-purpurea*. It was a product of Jackman's. Moore and Jackman write of it as being 'believed to be a natural hybrid between *C. flammula* and *C. viticella*, the seeds of the former from which it was raised having been grown in contiguity to some of the varieties of the latter species. In its vigorous growth, its profusion of flowers, its strongly-marked hawthorn scent, and its late summer and autumn-flowering habit, it partakes strongly of its mother-parent.' The flowers are, indeed, cruciform like flammula's but rather larger, 4 to 5cm across, and each sepal is purple round the margin shading to white in the centre. I was unkind to this clematis in my last book, dismissing it as 'grubby and ineffective', but I have to eat my words after seeing a well-grown and mature specimen in a Berkshire garden. It grows well up north. Even in the poor summer of 1987, Tom's two-year-old plant growing over a viburnum gave a good account of itself, flowering profusely from mid-August to October. He suggests that it needs a fairly rich soil. When I tried to grow it on a scruffy piece of yew hedging, it liked the position no better than the yew did. Crosses between *C. flammula* and *C. viticella* should be known as *C. × triternata*, so this clone becomes *C. × triternata* 'Rubro-marginata'.

It is unfortunate that *C. maximowicziana* is the correct name of the species more widely known and grown as *C. paniculata*. It comes from Japan (introduced 1796) and has long been popular in American gardens. Even in France it gives a good account of itself, flowering from late August to October, but it needs greater summer heat than we can generally stoke up. In most years, flowering in England starts too late and is beaten by the weather. However, Tom tells me of Harry Caddick's plant at Warrington, near Manchester, where even in a poor season the plant was covered with blossom on 4 October. This suggests that the flowering time of the species varies according to the strain being grown. The flowers, like flammula's, are carried in 'axillary panicles through most of the length of every trail, thus making up long flowering garlands' (*Revue Horticole*, 1902). There is the same flammula hawthorn scent. It would be worth having a stab at this species against a hot south wall in the south-east of England.

C. brachiata is similarly late-flowering. I was given this species by Maurice Mason in 1974, and Romke van de Kaa, who was gardening with me here at Dixter, drew my attention one October's day in 1975 to the fact that it was in full flower. As my eyes are usually cast modestly down upon the ground or else, at this season, focused on the sky in search of a few reassuring late swallows or martins, I had missed it. Summer and autumn had both been excellent and the blossom was in mint condition but could, I suppose, easily fail in a normal year.

This is a South African species, perhaps the only one such cultivated in Britain. Its leaves, at flowering time, are tatty, but one is unaware of them because of the profuse garlands of cruciform flowers, each 4cm across, green-tinted white with prominent stamens. There is a faint but pleasing scent. Others declare that the scent is quite strong, but I see no reason to believe that their sense of smell is more acute than mine; they must be showing off.

Romke thought that I should take it up to the next RHS Show on 28 October, and we duly set about preparing our material on the previous afternoon. We could not have laid a corpse out with greater solicitude. The clematis grows on an old espalier pear in an open and exposed part of the garden. We used all the flowering material and carefully detached each trail from its support, then re-attached them with clips on to stiff red branches of dogwood, *Cornus alba*, whose leaves had just been shed. First I had taken the precaution of boiling the cut clematis stems for 50 seconds. The reconstituted plant was next plunged in a deep florist's bucket for the night, in my porch. We stood back to admire. 'Fantastic' was Romke's reaction, and I must say it did look good but not showy. I doubted whether the committee would be impressed. But they were and it got an AM the next day as 'a tender flowering climber'. I rather resent that word 'tender', seeing that I could not have given my plant a colder position, and it lasted a good ten years, but I suppose they had to play safe and it is true that I lost it following the hard winter of 1985.

HERBACEOUS CLEMATIS
AND NON-CLIMBING SHRUBS

C. × *aromatica*

C. douglasii var. *scottii* and cv 'Rosea' 'Edward Prichard'

C. × *eriostemon*

C. heracleifolia and cvs 'Campanile', 'Crépuscule', 'Côte d'Azur'

C. heracleifolia var. *davidiana* and cv 'Wyevale'

C. integrifolia and cvs 'Hendersonii', 'Olgae', 'Rosea'

C. × *jouiniana* and cv 'Praecox' 'Mrs Robert Brydon'

C. recta and cv 'Purpurea'

C. songarica

C. stans

We can progress easily from the last group to this by way of *C.* × *aromatica*, whose parents are credibly presumed to be *C. flammula* (whence the scent) and *C. integrifolia* (whence the flower and habit). It has often been known as *C. coerulea odorata* and was described in the *Revue Horticole* of 1877, so probably originated in France about that date. Moore and Jackman wrote of it with enthusiasm as a continuous and abundant bloomer, from early July till autumn, so floriferous 'that the plants have the appearance of huge bouquets'. The flowers have 'a strong and delicious fragrance, resembling that of hawthorn'. It is semi-herbaceous and non-climbing, dying back to a woody base in winter, and as it grows 4–6ft high it should, they say, be fastened to some kind of support. The flowers are about 4cm across with four violet-coloured sepals that look narrow as their margins roll back. The button-like eye of stamens is in contrasting white.

If the reader feels that this is something he must have, that's how I felt, and I eventually got a piece from John Treasure's plant (it is not all that easily propagated). Alas, it seems to have lost a great deal in stamina during the century since its birth (well, don't we all?). My plant is a wee, frail little old thing. The scent is still there in the few flowers it opens but the plant, now eight or ten years old, has a distressing not-long-for-this-world look. Incidentally, Boucher et Mottet in *Les Clématites* (a disappointing little book on the whole) write of the flowers of *C.* × *aromatica* that they 'exhalent un agréeable parfum d'Héliotrope très suave', a description with a caress.

C. integrifolia had better be considered next. It is a herbaceous species from S. E. Europe and is sometimes classified as a member of the otherwise American section *Viornae*. But its sepals are not connate, as theirs are, and its foliage, devoid of any climbing properties and undivided, is utterly different. As a garden plant it still has its place. Growing 1m tall, each clumpy plant makes a forest of young stems from the crown in spring, and each stem terminates in one nodding indigo-blue flower of a somewhat leathery substance. If there are plants around it that will support its stems in their lower regions, that is ideal; otherwise you must give it discreet and sympathetic support yourself. Don't truss it to a stake.

The cultivar called 'Olgae' is about the same colour and size, but the sepals are even more attractively twisted and the flowers have some scent. Another cultivar, 'Hendersonii', received an AM when shown at the Chelsea Flower Show on 25 May 1965. It is easily brought on to flower for this show although it naturally has a midsummer season. The plant is usually less than 1m tall and the flowers are larger, bolder, more effective than in the unimproved species. They hang with wide-spreading sepals so that their total spread is 7cm and they are dusky indigo-blue, paler at the margins which undulate slightly. Although there is only one bloom to each stem, it stands well above (15–20cm) the foliage. 'Rosea', in a good form, is a deep lilac, but inferior impostors do exist. There are several more worthwhile-looking seedlings of *C. integrifolia*, some pinkish and some with quite sizeable flowers. A white clone is around but I have not yet come across it.

C. integrifolia 'Hendersonii' must not be confused with *C.* × *eriostemon*, among whose several synonyms *C. hendersonii* is one. *C.* × *eriostemon*, as already mentioned, was the earliest hybrid, the cross arising in 1835, probably between *C. integrifolia* and *C. viticella*. It is a non-climbing plant, its weak though woody stems up to 3m tall. The flowers take after *C. integrifolia* but are not really sufficiently numerous or impressive for the size of the plant. After growing it in my mixed border for a number of years, I decided that it made too much fuss with too little result.

In all the early literature, *C. heracleifolia* was referred to as *C. tubulosa*. There is a clematis group still referred to as Section *Tubulosae*, of which this is horticulturally the most important species with *C. stans* as also-ran. Their sepals are joined at the base in a tube and the flower form, colouring and size are very much that of a hyacinth's.

C. heracleifolia is a common sub-shrubby clematis and variable too. There are a number of clones derived directly from it; others from crosses with *C. stans*. Lemoine of Nancy was particularly keen on improving these small-flowered herbaceous or sub-shrubby species, and the greatest number of crosses were made by that nursery. 'Campanile' is one such that is still with us (by name anyway, and the early descriptions more or less fit). It received an AM in 1917 when shown by Messrs Paul of Cheshunt. They described it as herbaceous, but today's plant makes thick woody stems in course of time. 'Lilac blue flowers with white centres', as described in *Le Jardin* at that time fits well. 'Côte d'Azur' is said to have deep azure-blue flowers, but was, as I acquired it, identical with 'Campanile'. Once again there has been a mix-up.

C. heracleifolia is a coarse clematis on account of its voluminous leaves which tend to be out of all proportion to the size and quantity of blossom. However, a well-flowered specimen is worth looking at. This species comes from central and north China and was introduced in 1837. *C. heracleifolia* var. *davidiana* was collected by Père David near Peking and was sent by him to Paris in 1863. Botanically, the most obvious features distinguishing this from the type are (1) that *C. heracleifolia* is monoecious (male and female organs on different flowers but on the same plant) while var. *davidiana* is dioecious (the sexes on different plants); (2) that the sepals roll back at the tips in *C. heracleifolia* but merely spread outwards without rolling back in the variety. I go along with the first distinction; the second is less marked. *Davidiana* rolls back quite a lot, especially in the *newly* opened flower, funnily enough. Writing as a gardener and not as a botanist, the main distinctions as I see them in the plants we have in cultivation are that *C. heracleifolia*, the type, is sub-shrubby, the annual growths dying back to a noticeably woody and permanent framework; while David's variety is truly herbaceous. The latter spreads, moreover, like many herbaceous plants, by means of underground rhizomatous shoots into a clump that is readily divisible; not so *C. heracleifolia*. Furthermore, var. *davidiana* is strongly scented – a somewhat disconcerting hair-oil scent – whereas the type-plant hasn't a vestige of scent.

The flowers in this species are carried in dense axillary clusters and are larger in var. *davidiana* than in the type, but largest in *C.h. davidiana* 'Wyevale', an excellent clone in which the colouring is a strong blue throughout the flower, which also has attractively crimped sepal margins. It won an AM when I showed it on 7 September 1976.

Among Lemoine's many *heracleifolia* × *stans* crosses, 'Crépuscule' is one that has survived, but something has re-entered growers' lists as 'Bonstedtii Crépuscule'. It has an altogether smoother leaf and a more compact habit than *C. heracleifolia*. 'Mrs Robert Brydon', raised in the USA, may be pure *C. heracleifolia* or it may be a *C. × jouiniana* cultivar: a strong grower to 2m (it needs good support) with sprays of pale off-white blossom faintly blue-tinged. The flowers open wide, curling back at the tips and have no tube (which suggests vitalba blood). The white stamens are prominent. This is mildly agreeable but of a too dirty colouring. If I haven't sufficiently damned with faint praise let me add, unequivocally, that it is a poor thing.

So is *C. stans*. This is a floppy 1–1·3m herbaceous species carrying flowers of a spitefully non-contributory off-white, skimmed-milk colouring. 'Of little importance,' wrote Moore and Jackman. Hear! Hear! *C. stans* 'Lavalei' has entered the Continental growers' lists of late, but I have had no experience of it.

Now for an interesting and worthwhile clematis, the hybrid *C. × jouiniana*, widely and erroneously distributed in former times as *C. grata*, which, however, is a genuine (if undistinguished) species in its own right related to *C. vitalba*. Jouiniana itself is a cross between *C. heracleifolia* var. *davidiana* and *C. vitalba*; the only cross, as far as I know, in which our native clematis has been used as one parent. It was made (or should I say the hybrid occurred?) about the end of the last century and was named after E. Jouin, manager of the Simon-Louis nurseries at Metz.

Seeing that both its parents wear coarse, obtrusive leaves, for which I have duly castigated them, you might think that the hybrid would be similarly burdened. Indeed, its large pinnate foliage is no asset, but the flowers are so numerous that, once in full spate, the leaves practically disappear. These flowers make wide-open, off-white stars, having a bluish tinge that is strongest on the sepals' underside. There is no perceptible scent but the blossom is popular with butterflies, as is davidiana's.

The habit of the plant is to make a basic, permanent woody frame up to 1m high and then long annual growths to 3m that flower as a compound panicle along almost their entire length. Given wire supports, *C. × jouiniana* can hoist itself up the walls of a two-storey building, although it has no climbing devices. I use it as a border feature and give it height by tying the old stems to a post; from this the young growths spray outwards and downwards in all directions. But *C. × jouiniana* is also one of the best clematis to use as ground cover in an open situation or even in thin woodland, and it can mask old tree stumps.

Do not, however, fall into the trap of thinking that any old soil will give results. Tough though it is, *C. × jouiniana* is a gross feeder and allowances should be made for this when establishing a new plant in rough or uncultivated ground.

The original type-plant of this hybrid does not start flowering till September. Usefully late, you might think, but in fact this very late start does unnecessarily curtail its season. The cultivar called 'Praecox' should be chosen in preference almost every time, because its season starts in July and thereafter continues without abatement to the end of September or further.

'Edward Prichard' is the interesting result of a unique cross between *C. heracleifolia* var. *davidiana* and *C. recta*. It was made by Mr Russell V. Prichard in Australia, and the firm of Maurice Prichard (lately of Christchurch, Hants, but now defunct, alas) gained an AM for it on 29 August 1950. Its herbaceous growth is thin and weak to 1m, and it carries sprays of pale mauve, strongly scented cruciform flowers (the sepals narrow for their length) in late summer and autumn. It never does much in my mixed border. I think capsids eat out its young shoot tips and buds. But it can be a really charming clematis. Perhaps it is not quite robust enough to cope with the hurly burly of mixed border life.

And so to *C. recta*, at one time called *C. erecta*, which means the same thing and seems to me the silliest name that could have been chosen for a clematis that is so floppy by nature that this aspect of its habit is the most difficult to cope with in the garden. You can embattle its young growth with 2m-long pea-sticks at an early age, but this is expensive and tedious and looks as though you'd planted a witch's broom. Or you can let it flop, but that's not much fun for the neighbours on whom it collapses. There are strains of this clematis that grow only 1m high and are more manageable, but it is generally twice as tall. As it self-sows extremely readily, there are innumerable clones in circulation.

I shouldn't bother about *C. recta* straight. You might just as well go for 'Purpurea', which flowers similarly for three weeks at midsummer with a delightful foam of small white cruciform blossom (excellent for flower arrangements if you boil the stems first), but additionally is sumptuous in spring when the young shoots are coloured rich purple. This really does take the eye, but the colouring fades to dark green as the foliage expands. Even 'Purpurea' is not a fixed clone, as it has often been propagated from seed. Hilda Davenport-Jones told me that she treated it this way, for sale, discarding the green seedlings and selling the purple. That's too hit-and-miss a method for my liking. You can divide old clumps or you can take cuttings of young shoots in early spring when only 5cm long, detaching them right at the base where they arise from the crown, or you can make internodal cuttings in the normal way.

As a garden plant I must admit to finding *C. recta* tiresome, by and large. Lemoine brought out a double form, 'Plena' or 'Flore Pleno', which won an FCC when shown by Messrs Paul to the RHS in July 1890. It is mouth-wateringly described by Moore and Jackman as having 'pure white flowers which differ from those of the type in being fully double, like the silvery, button-like blossoms of *Ranunculus aconitifolius plenus*', to which, of course, it is related. I bought a plant under this name from Thompson & Morgan years ago, but it was just the ordinary single-flowered species, in the event. A sad loss.

C. songarica is an interesting autumn-flowering species from Siberia (introduced 1880), of a sub-shrubby habit reaching to 2m against a wall. The smooth green leaves are entire; the flowers, in terminal clusters of about thirty, are pure white and starry, about 3·5cm across, having five sepals usually. They are hawthorn- or meadowsweet-scented. There is a specimen in the car park (an old walled garden) at the National Pinetum, Bedgbury, Kent, which they had from Kew: near the entrance to the pinetum itself. Most of the clematis on these walls are vitalbas!

Finally *scottii*, a variety of the herbaceous loner *C. douglasii*, discovered by Douglas himself in the Rocky Mountains of the USA. When suited, this variety makes a handsome clump to 1m tall at flowering, in early summer, with somewhat glaucous pinnate foliage and solitary, nodding bell-shaped flowers, 4cm long, opened at the mouth, blue, darker within than without. 'Rosea' is a good shade of pink, not muddy.

This clematis is easily raised from seed but seems reluctant to make a good plant thereafter – with me, at any rate. It grows well in John Treasure's garden at Burford (Shropshire). Tom has tried it with seed from John's plant. He says that it has a nasty habit of collapsing to ground level in mid-season but is rarely killed outright. John advises that only strong plants should be planted out. Keep weaker ones in a pot to build up their strength.

Chapter 7
HOW TO GROW THEM

Clematis have a reputation for being difficult. I was sent for review a *Handbook of Easy Garden Plants,* intended to recommend to beginners a range of trouble-free plants that will allow the housewife to lounge in a deckchair rather than take up a gardening tool. Naturally I looked up clematis first. The only two allowed by the author were *C. montana* and *C. tangutica.* After describing them she went on to say: 'The large-flowered hybrids are less easy to establish than the above species, and are decidedly fussy, so have regretfully been excluded from this book.' End of clematis.

I admit I'm no deckchair gardener. The pleasure you get out of gardening is in proportion to the mental and physical effort you put into it. That's the way it works with me, anyway, but if you're a reluctant gardener and have been pushed into gardening against your inclinations, then what about the large-flowered clematis for you? Are they really so fussy that you can never be recommended to plant one? Quite frankly, I find you such a bore that I don't care whether you do or you don't. I can say that here without giving offence, because you certainly won't be reading this book. Only the converted clematis fan will be doing that. However, I must point out that from the north of Scotland down to the south-west of Cornwall any motorist with his eye off the road will see thousands upon thousands of purple *jackmanii* clematis flowering on innumerable house walls and porches every July and August. Don't tell me that they thrive only by the sweat of each owner's brow. On the contrary, ninety-nine out of a hundred of them were planted with no more thought than is given to drinking a glass of water.

In a specialist book on the clematis, I am naturally going to give a good many details and instructions to help the reader get the best from his plants, but this wealth of detail should never be allowed to obscure the fact that if you simply dig a small hole, place the clematis in it, fill in the hole, turn your back on it and make for the nearest deckchair, that clematis stands an excellent chance not merely of survival but of flourishing in a manner that does you credit such as you most certainly do not deserve.

I have in the past weeks, during a fantastically mild January, planted about two dozen clematis against old, unproductive espalier pears, (they'll be even less productive now). I am sure I didn't average ten minutes of my time on each of them; five would be nearer the mark. But I should add that the soil has long been cultivated and is, on the whole, in good heart, as the saying goes. Only here and there was I battling with yellow, cheesy clay that had to be replaced with soil-lightening grit.

When to Plant

The question that a nurseryman is asked more frequently than any other is 'When is the right time to plant a clematis?' It is usually asked by someone who wishes to defer the issue and doesn't feel ready to cope. He does not wish to be told that now is the moment, and yet, as clematis are nearly always pot-grown, they can, without disturbance or damage to their roots, be planted at any season. Obviously not if the ground is frozen solid, but you can keep the plant by you in a cold shed until the frost relents. Obviously not if the ground is flooded, but in that case the situation is unsuitable anyway and steps must first be taken to drain it adequately. If the ground is bone dry, you can, unless there are water restrictions in force, water it until it reaches saturation point; wait for it to dry a little and then plant your clematis. But the point to make here is that there is no close season for clematis planting and plants will be available from garden centres throughout the year.

Mail orders will be dispatched between late September and the end of March, generally speaking, though they may continue through the spring as late as early June. Clematis are seldom dispatched to customers during the summer, (a) because being in full growth they are more liable to damage by rough handling or delays in transit; they may dry out or get too hot in their parcel during a heat wave; (b) because a nurseryman has many other calls on his time in the summer and wants to use his labour force then on other tasks than packing and dispatching.

A clematis that is being committed to public transport will travel best when temperatures are low (they are seldom too low) and the plant is dormant so that it is carrying no soft and easily damaged growth.

It sometimes happens that the clematis you receive are far weaker than they should be, in which case you will find little evidence of root development when you turn it out of the pot. It is extremely hazardous to commit these straight to the garden. They will need to be treated as hospital cases. Pot them up and keep them under your eye until their strength has clearly built up and the pot is full of roots.

Some gardeners have definite preferences as between autumn and spring planting, having discovered from experience that one or the other has been followed by losses. More often than not this is *post hoc* and not *propter hoc*. Most clematis being fully hardy in Britain, winter losses can seldom be ascribed to the cold killing a not yet established plant. Spring planting may often be followed by drought, and if you or your mind are not on the spot, an unestablished plant could easily fold up; but forewarned is forearmed.

In sum, my recommendation is to buy any common clematis when you are ready for them and plant them forthwith and to snap up the uncommon kinds that are frequently out of stock whenever you can lay hands on them. They can easily be heeled in, in a spare bit of ground, till you have made their permanent quarters ready, even though that may be months later. Desirable clematis that are often in short supply include 'Alba Luxurians', *C. armandii,* 'Beauty of Worcester', 'Countess of Lovelace', *C.* × *durandii*, 'Etoile Rose', *C. florida* 'Sieboldiana', 'Gravetye Beauty, 'Jackmanii Rubra', 'Perle d'Azur', *C. rehderiana,* 'Royal Velours', 'Sir Trevor Lawrence' and *C. texensis.*

Soil and Moisture

To the question 'What do clematis like?' the short answer is 'Everything you can give them.' And I find it helpful to add that if clematis are fed as most gardeners feed their roses, they will be well content. Roses, in fact, are often overfed, but this is unlikely to happen with clematis provided ample moisture is there.

But first to consider what soils clematis like and to what extent you can or should try to modify your soil as you find it.

The commonest misconception is that clematis need providing with lime if none is present in the soil. Let me assure you straight away that this is quite unnecessary (unless your soil is so acid that it needs extra lime for everything you grow, and this is rare in British gardens). There is no need to go scrounging with your bucket for mortar rubble out of old buildings. All the same, you will still read in many books and articles on clematis that lime is a prerequisite. I have no doubt that this notion arose from the fact that our one native clematis, *C. vitalba,* is most often found wild on chalk or limestone. One swallow doesn't make a summer. *C. vitalba* has sired only one hybrid grown in gardens. And anyway itself will grow perfectly well in acid soil if you ask it to. So will all the rest. I remember Miles Hadfield telling of the large collection of clematis in his parents' garden in Birmingham (and indeed I remember an ancient *C. armandii* on a wall of their house), where the soil was excessively acid on account of being impregnated by sulphurous industrial fumes. But the clematis flourished exceedingly. This may also, I suspect, be linked with the fact that they got less wilt disease, because the sulphur dioxide in the air acts as a fungicide. It is well known that potatoes do not get blight nor roses black spot – both fungal diseases – in areas of industrial pollution and smog. I am not a town gardener myself but my London friends seem to do clematis very well and with the minimum of trouble from wilt.

Walter Pennell thought differently from me on this matter. He wished to go on record as believing that chalk or lime in the soil where clematis grow materially reduces the incidence of wilt. His own garden was on limy soil and he had little or no trouble from this disease. If he was right, we have yet to be given a clue as to the reason. It used to be the practice to whitewash the trunks and branches of fruit trees with lime, in the interests of orchard hygiene. No one does it now. I presume that the whiteness of the lime gave the grower a notion of cleanliness and hence of protection against disease.

The disadvantages of adding lime to the soil where you grow a mixed collection of plants are clear. It won't hurt the clematis but it will make rhododendrons, camellias and most heathers difficult or impossible to grow, all of them excellent host plants for clematis to clamber over, and there are many more that are similarly calcifuge.

The most difficult soil on which to make a success of any but the toughest clematis is a hungry sand. Such are the Folkestone and Bagshot sands; poor soils that dry out. Gertrude Jekyll gardened on this kind of soil at Godalming, in Surrey, and she commented on the long years that it took to establish a *C.* × *jackmanii*. 'In my garden it is difficult to get the Clematis to grow at all,' she wrote. 'But good gardening means patience and dogged determination. There must be many failures and losses, but by always pushing on there will also be the reward of success. Those who do not know are apt to think that hardy flower gardening of the best kind is easy. It is not easy at all. It has taken me half a life-time merely to find out what is best worth doing, and a good slice out of another half to puzzle out the ways of doing it.'

It is the lack of moisture that makes a sandy soil difficult. Thin soils overlying chalk can be tricky for the same reason. But some limy soils are surprisingly moist even in times of drought. Provided their drainage is adequate, soils overlying clay are among the best for clematis, but even these can dry out at times and this must be guarded against.

Remember that clematis are frequently planted near walls. The wall itself laps up a large share of any moisture that's going, and its physical presence also prevents a good deal of the rain that falls from reaching the soil at its feet. So clematis near walls need extra special consideration, on the water front.

There are two ways in which the gardener should, concurrently, approach this problem. The first is by adding water to the soil himself, the second is by facilitating retention of water in the soil once it is there.

The best kind of soil from every point of view is that which is rich in humus: decayed vegetable matter like leaf mould, peat, farmyard manure, garden compost, straw, spent hops, sewage sludge, road sweepings, ground bark as a forestry by-product, deep litter chicken manure based on sawdust or wood shavings. When I am asked what I recommend as a soil conditioner for clematis I think of all these things, but which to recommend depends so much on the individual's locality and what he can acquire or make most conveniently at the lowest cost.

Some in the list I've just given have feed value, others (like peat) are purely or mainly soil conditioners in the form of humus, which will hold moisture, aerate the soil physically and also encourage the soil fauna of which earthworms are the most noteworthy. So you want to add humus to the soil before you plant your clematis and you sub-sequently want to add it as a surface dressing or mulch, topping it up for as long as the clematis lives. Mulches must be applied when the soil is well saturated, which usually means in winter, and they will then, if thick enough, keep the soil beneath cool and damp even through quite prolonged droughts. Indeed, the old and oft-repeated recommendation that clematis need shading at the roots will no longer apply. John Treasure has grown them in many open situations, mulching heavily, without the need for extra shade or protection.

If you are obliged to plant a clematis in a quite small aperture surrounded by paving, there is little opportunity for adding humus and mulching. As long as the paving slabs are not cemented together, however, they do keep the soil underneath them remarkably cool and moist so the clematis may be better off than you'd think. Scope for feeding will be restricted, and if the soil is poor anyway you may have to settle for strong clematis like montanas, macropetalas, alpinas and viticellas that are not too fussy.

Now as to actually watering your clematis yourself. When you have decided to do this, the great thing is to give them enough so that it penetrates deep into the soil and reaches all the plant's roots. If there is a surface mulch you won't have to water so often, but when you do you'll have to be even more thorough than usual, because the water has to saturate the mulch before it can reach the soil underneath. A noticeable benefit from the mulch is that it will prevent water from running off and away, as it is so liable to do when it falls on a hard, impermeable soil surface. That, of course, is useless and must at all costs be avoided. A slow application in fine droplets will ensure that the water is absorbed where you want it.

As to quantities, think of them in this way. The growing plant (not only clematis) needs 25mm (i.e. an inch) of rain or other water every ten days. Your watering needs to make good the deficiency in natural rainfall in each ten-day period. Twenty-five mm of water is equivalent to 25 litres to the square metre (or $4\frac{1}{2}$ gallons to the square yard). This is a great deal more than most people imagine a plant needs at a watering, especially if they are putting it on from a watering can.

Many gardeners (indulging in the lazy streak that lurks in them) object to watering on the alleged grounds that once you start you have got to go on. I cannot myself see that this is an objection, except for the bother of it. Of course you should go on, for as long as the need is there. Even so, one thorough watering in a dry spell is better than none, even if you don't keep it up, just as one good day's rain in the middle of a dry spell would be better than none.

If you live in a district where restrictions on water use are frequently imposed during times of drought, then it is more important to be liberal with the humus and to keep your mulches thick at all seasons. If mulches build up too high around the stems of some shrubs, they can induce neck rot, a horrid complaint, as you can quickly appreciate simply by murmuring those two words over and over again to yourself: neck rot, neck rot, neck rot. Terminal neck rot, indeed. But you can heave a sigh: it luckily works the opposite way with clematis. They positively enjoy having the goodies piled around their necks and react by rooting into them. Also, in many cases, by throwing up strong shoots from below mulch level.

Before continuing with details of planting preparations and procedures, I might as well finish off the general subject of feeding, as we have already been teetering on the brink of it for some time.

Although I said at the start of the chapter that clematis are greedy, that does not mean they need better treatment than the plants which I hope you are growing all around them. Under garden conditions, nearly all plants need and deserve generous treatment.

Some of the forms of humus that I have been reviewing also have considerable feed value: farmyard manure, for instance, especially if it has been stacked under cover so that most of its nutrients have been retained, not washed away by rain. Deep litter chicken manure is exceedingly rich because chicken droppings are anyway strong and also because practically nothing is lost in the making, which is automatically under cover. When you empty the deep litter house you must keep the litter in a dry place, and it may want turning once or twice before it loses its strong ammoniacal smell. Until then it will not be sufficiently rotted to be applied safely to growing plants. I use a lot of deep litter manure myself, as our hens make it, but a little goes a long way. It only needs thin spreading.

Grass mowings, on the other hand, can be spread thickly – say 10cm deep. They won't heat at that depth but they will, in the course of rotting, take nitrogen *from* the soil rather than contribute nitrogen *to* the soil. And that goes for any undecomposed vegetable material that you may apply as a mulch. Straw, for instance, or farmyard manure in which the straw has not yet broken into short pieces and the animal smell is still obvious; or bark, or sawdust.

Such materials can be applied safely as a surface mulch to woody plants like clematis, especially in the dormant season, but as soon as growth recommences in the spring you should add a surface dressing of nitrogenous fertilizer to make good what is being lost from the soil to the mulch in the course of decomposition. Your mulch will also break down much more quickly and become valuable as a manure itself if you add nitrogen. Sulphate of ammonia or nitro-chalk will do very well.

Various new formulations have become available in garden centres. The controlled-release fertilizers (Ficote/Osmocote) are used extensively by growers, and you will see these as blue or amber capsules (the size of a peppercorn) in the compost when you turn the plant out of its container. The amber variety can look rather sinister, inducing in the uninformed a worry that they are the eggs of some large and nasty insect pest. They are designed to release nutrients at a controlled rate over a longish period. The proprietary liquid feeds can give good results but need to be applied every few weeks. Dry organic fertilizers can be effective throughout the season, following a single, large application in the spring.

With all fertilizers it is essential to look carefully at the formulation. This is expressed as a ratio – nitrogen (N): phosphorus (P): potassium (K), (i.e. N:P:K) – and if you are intending to apply this as a general feed as I recommend you should in spring and summer, the ratio should be more or less equal. If a plant is weak it will *not* respond to heavy feeding. Make sure that it is already growing healthily before you apply fertilizer.

Planting Preparations

Planting preparations on a good soil will take two minutes. You will simply fork in some humus to a fork's depth. Earthworms will carry it down as much lower as it needs to go. If the humus you add is peat or some other non-nutritive material, then you should also include a couple of handfuls of John Innes Base Fertilizer, which contains nitrogen, phosphorus and potassium, and will release these nutrients over a longish period, not all at once. But I don't add this where my soil is really good.

On nasty soil, adequate preparations may take half an hour for each plant. And that is in addition to the adequate drainage provisions which are a *sine qua non* for all gardening except in bog and pond. If you dig out a large hole the sides of which are neat clay, the fact that you fill in with delicious John Innes No. 3 Potting Compost will not prevent your hole from becoming a sump, unless tile drains have been laid near at hand.

Assuming that you are on clay but have solved the drainage problem, I recommend (and this is what I do myself in the grottier parts of my garden) setting the reasonably good topsoil on one side of an area 0·7m (2ft) square. Then, down to a depth of 0·5m, extract and remove the yellow cheese. Replace the top soil (which will now be at a lower level) and add a similar bulk of horticultural grit. Also some humus and organic manure as already discussed. Mix these together with a long-tined garden fork. Top up with more topsoil filched from elsewhere or, better still if you can get it, with the aforementioned John Innes No. 3 compost.

Horticultural grit is available all over the country but you may have to show initiative to locate a source. In my part of the south-east, in Kent, Sussex and Surrey, it is available from M. P. Harris, the builders' merchants. If you order a load at a time, they will deliver. This sounds like a lot of grit but it has many uses in the garden: to mix as 'sand' in John Innes composts, to bond paving stones, to spread on much trodden dirt paths, to lighten the ground where you intend planting a lavender hedge, to use as a surface dressing to keep down weeds or to make them easy to pull out. If you only want a small quantity you can take your bag or sack to the nearest depot.

The grit around here is ground up shingle or 'beach' as we know it locally. What you see in heaps by the roadside, 'Grit for Icy Roads', is very similar. In other parts of the country, what you want may be sold by different merchants and come from a different source. Cal-Val market it as 'coarse grit' and offer it in bags of 5kg and 25kg. This is handy but expensive if you need large quantities.

Back to our planting site. Similar preparations to those described will be worth making in any unpleasant soil, whether sandy, gravelly, rocky, chalky or whatever. The lighter the sub-soil the greater the bulk of humus that should be added. This is where peat is so valuable, and it is marvellously easy to handle, but it has become fearfully expensive latterly.

Planting

Although the actual planting of a clematis takes a matter of seconds, there is quite a lot of jabber apropos. First, as to the type of plant you have bought and the season at which you're planting it.

Most young clematis will be pot grown, but some will be acquired in plastic bags and be quite devoid of soil. No harm in that, especially if they are dormant. Dig a hole large enough to accommodate the roots spread out – not a small hole into which the roots have to be bunched up.

Pot-grown clematis have the advantage of needing very little root disturbance in the process of planting. Clematis root systems vary a lot in appearance: that is, in their thickness and strength. The majority of the hybrids have thick, tough roots distinctly reminiscent of leather bootlaces. Parkinson, in 1629, wrote of them in his *Paradisus* as 'a bundell of brownish yellow strong strings'. If they were cramped in their pot it is a good idea, before planting, to unwind and release them a little one from another, working from the bottom of the root ball. In this way you help them to make quick contact with the new soil into which you are about to plant them. So tough are these roots that your handling of them will do practically no damage. But I shouldn't mess around in this way if the plant is green and growing strongly at the time you are planting it. In that case (say it's spring or summer) turn it out of its container (even if it's a perishable paper pot) as gently as possible; place it in its hole and then pour in a gallon – sorry, I mean pour in 4·546 litres of water, quickly. As this is draining away, quickly fill in with top soil.

Some clematis have very fine, easily damaged roots and you should not manhandle these at all at any season. The alpinas and macropetalas come to mind as examples, but worst of all is *C. orientalis*. If you are a bit careless, the ball of soil and roots can break clean away from the stem while you're handling them, and the plant will certainly die. Don't thereupon write an irate and self-righteous letter to the nurseryman telling him that his clematis had no roots. It did have roots, but quite unlike those on the hybrids that you possibly bought at the same time. And you bungled. Of course if it was a mail order, the accident may have happened through inadequate packing and rough handling in transit.

If you receive your clematis in autumn or winter, they may be hung with dead leaves and look completely dead. The trouble with the majority of large-flowered hybrids is that they cling to their old leaves at the end of the growing season, since they have not the knack (as has *C. montana,* for instance) of shedding them. This certainly looks terrible but it means nothing. I do think, though, that it is wise of the nurseryman to pick off these dead leaves before dispatching plants, because a naked plant looks so much better than a scarecrow. Also, some of the dead leaves might be carrying fungus spores and act as a source of infection by wilt disease at a later stage.

Another trap for the unwary. Young clematis are generally tied to canes. If the cane was not placed fairly centrally in the pot or if it was rather too heavy for the strength of the roots in their ball of soil, then, when you turn the clematis upside down so as to knock or jerk it out of its pot, the cane may take on a life of its own in a most disconcerting fashion. It may twist sideways and kink or even break the clematis stem near the base. That is not necessarily fatal by any means but it may, with the fine-rooted kinds I've just been describing, bring about a severing of roots from stem, which is the end.

If you receive the clematis in a paper pot or in a polybag (a black polythene bag in which it was grown), you can cut and peel the containing material away from the ball of roots while the latter are sitting in their hole, ready for planting, and there'll be no need to turn anything upside down. But if the plant is being turned out of a rigid pot, hold it vertically upside down (not obliquely) before you tap the pot rim to extract its contents, and stretch the fingers of your left hand (the hand that's going to support the ball of soil) round both sides of the cane (two on one side, two on the other), so that they cover the whole top surface of the ball and grip the cane at the same time. It is just at the moment of extraction that nasty things are liable to happen. Righting the plant is not nearly so tricky. After ruining a dozen or two you'll find you've got the knack and wonder what all the fuss was about.

Two other pre-planting points are perhaps worth making. If the ball of soil is dry on arrival, as it may well be after a journey under the aegis of national transport, drop it, container and all, into a bucket of water. Half-an-hour will be long enough if the plant was grown in a soil compost like the John Innes formula. Nowadays it is more likely to have been grown in a peat and sand compost without soil, and it should then be soaked overnight, as dry peat is very reluctant to take up moisture.

Second point: if the plant was grown in one of these soilless composts and you are not ready to plant it on arrival, remember that it will soon run out of nutrients and should be given liquid feeds, say once a week in the growing season, if the delay in planting is going to be protracted.

When planting, firm the plant in so that the soil round and below it isn't too soft and spongy and get it placed so that the lowest 2cm of the clematis stem is below the soil surface. Water in well if there is the slightest hint of the soil being dry.

The special problems of planting clematis in turf or in the proximity of tree, shrub or hedge roots have already been discussed in Chapter 3. If you must plant in turf, keep the grass back from the clematis stem by at least 0·7m. Water and feed extra-generously. If birds scratch your mulches on to the grass you'll have to net them in.

Aftercare

The pruning of young stock is described in the next chapter.

The main point to remember with a young clematis, even more than with an established one, is that it shall not go short of water. An old gardening friend, who has long been a clematis addict, recommended to me (years ago) knocking the bottom out of used claret bottles and then sinking them in the ground, bottom up, beside the newly planted clematis so that by watering through the bottle you made sure of getting down to their roots. In order to make things easy, he gave me a case of excellent claret which I drank with great pleasure but entirely forgot to use the bottles afterwards. I'm not sure that burgundy would come to the same thing. A snag we're running into today is that fewer bottles are being made with a kick-up, which is a perfect funnel when you're watering. It also facilitates knocking a hole in the bottom without breaking the entire bottle.

It is a fact that puppies are very much given to cutting their second teeth on clematis stems. With a dreamy, far-away look in their eyes, they chomp and chew right through even quite thick trunks. Humans themselves have been known to tear through young clematis stems when they intended to be doing something quite different. You can resort to a circle of wire netting. I don't like the look of the stuff myself and would rather risk disaster, but then I've always got young clematis coming on that I want to find places for. A plastic collar could be used (as it is for newly planted trees) to protect young shoots and stems from damage (see also p. 156).

Once a clematis is established you must resist any temptation to dig round it, as the roots reach right to the surface and spread far and wide beyond the trunk. The best way to ensure being good is to keep that surface mulch going. Wherever there's a mulch in the garden you won't want to dig – another point in its favour.

Chapter 8
PRUNING AND TRAINING

An unpruned clematis looks like a disembowelled mattress – a painful sight. From year to year its owner grows more despairing; his only comfort is that the birds are nesting in the tangle and perhaps not all of them are sparrows. Every time he bumps into this dropsical lump, his face, neck, clothes get sprayed with water. He loves clematis but he abhors this loathsome abortion. What is he to do?

The trouble with clematis is that they cannot all be pruned in the same way. For the busy man who has not the time to go into whys and wherefores, my kindest policy to save him reading further will be to lay down three basic pruning methods, one of which can be applied to almost any clematis he may grow.

Group A. Prune only if space is limited. In that case cut out all shoots that have flowered immediately after flowering. This method applies to the earliest, spring-flowering species, hybrids and cultivars, notably *C. alpina, C. macropetala, C. armandii, C. montana* and *C. chrysocoma.*

Group B. Cut out dead growth. Shorten all remaining vines to the first pair of strong buds. Time of pruning: February–March, when the buds referred to will already be plump and green. To this group belong all the large-flowered hybrids (including the doubles) that are expected to start their main flush of blossoming before mid-June. Examples: 'Nelly Moser', 'Lasurstern', 'Miss Bateman', 'Duchess of Edinburgh', 'Vyvyan Pennell', 'Mrs Cholmondeley'.

Group C. Cut all growth hard back to one metre or considerably less above ground level. Time: February–March. This will entail the removal of many healthy-looking green shoots but don't let that worry or deter you. The method applies to all later-flowering clematis that are required to make their main contribution after mid-June (perhaps as late as September, in some cases). Examples: *C. viticella, C. rehderiana, C. flammula, C. tangutica, C. × jackmanii,* 'Perle d'Azur', 'Royal Velours',

'Lady Betty Balfour', 'Duchess of Albany'.

In the alphabetical descriptive section at the end of this book, an A, B or C will be found against every clematis that can be attached to one of these groups. A few will obligingly slot into either Group B or C, allowing considerable latitude and indulgence for carelessness. Anyone wishing to work to a quick rule of thumb should thus be satisfied.

Now let us start at the beginning again and take an intelligent interest in how the different kinds of clematis are behaving and how our pruning is best calculated to take account of their behaviour.

The earliest clematis (Group A), including *C. cirrhosa* var. *balearica* (syn. *C. calycina*), have flowers that are 'borne directly on the last season's wood' (to borrow Mr Pennell's wording). The clusters of from 1 to 6 flowers and leaves that you get in the axil of each main leaf are really condensed shoots: so condensed, indeed, that they are stemless. In clematis flowering a little later, however (Group B), these axillary shoots do make a little extension growth and at least one pair of young leaves before you reach the terminal flower bud. 'Flowers borne on short growth from last season's wood' is the way Pennell's succinctly phrase it. 'Nelly Moser' and 'Marcel Moser' are typical here.

We next come to certain clematis that refuse to conform to one pattern. First they produce a flower at the end of each fairly short growth from last season's wood; then they make masses of extension growth on long, vigorous young shoots and proceed to flow along the last metre or so of this, carrying one or more blooms in the axil of each leaf. Typical here are 'Ville de Lyon', 'Ernest Markham', 'Hagley Hybrid', 'Marie Boisselot', 'Mrs Cholmondeley' and 'Duchess of Sutherland'.

Finally we reach the late, Group C, flowerers which do it all on the last metre or two of long, current season's shoots. Even here one may get the occasional early bloom off the old wood on an unpruned specimen. I have known 'Gipsy Queen' do this in June, although its main flowering is in August. The most prolific and long-flowering within this group carry many blooms in the axils of each node, not just one (as in 'Ville de Lyon'). The main axillary bloom will be supported by further flower buds which will open later and extend the flowering season, albeit with smaller blooms. Such are 'Comtesse de Bouchaud' and 'Perle d'Azur'.

I hope you get the overall picture. No new growth has to be made to enable the earliest-flowering clematis to perform, but the later you get in the season, the more growth has to be put on before any flower buds are formed until we reach 'Lady Betty Balfour', in September, flowering at the ends of 3 or 4 metre-long shoots of the current season.

The main point about pruning is to help each clematis to carry the maximum of flowers. To this end, of course, pruning must go hand in hand with watering and feeding.

Back to the early flowerers again. Group A includes some very vigorous clematis, and *C. montana* is far and away the most vigorous of the lot. They are set the task of masking eyesores like new garages quickly but, having done so, are then required to lay off it. But you can't explain that sort of rule or requirement to a plant and, not unexpectedly, they just keep on growing. This exposes them to the most damaging line in butchery. The irate owner, finding that he can no longer see out of his windows nor drive into his car port, gets cracking with his secateurs in the autumn, which is the gardener's natural tidying-up season. I am not saying that drastic action may not be necessary then, at the end of the growing season, but I do think that if the owner, given a reasonably phlegmatic temperament, realized that with every trail of new montana or macropetala growth that he was removing in the autumn or winter he was depriving himself of fifty or more blooms the next spring, he might well find that 'operation haircut' could be deferred until after the flowers had been enjoyed. Some time in May or early June in nearly all cases.

If pruning after flowering, in May, you can cut a clematis back to a framework of branches, leaving not a stitch of foliage on the whole plant. No harm will come. With the whole summer before it, your plant will start into growth again very shortly after its massacre and will be amply covered with strong young vines by the autumn. On these it will flower in the following spring.

On these early flowerers, then, we practise a kind of delayed pruning, and it is worth observing that many others of the earliest-flowering shrubs are treated in like manner, e.g. *Prunus triloba* and *P. tenella*, *Spiraea thunbergii* and *S.* × *arguta*, *Forsythia suspensa* and *Jasminum nudiflorum*. You can shorten the branches on all these after flowering because they have the summer before them in which to make next year's flowering wood.

Treated thus, these early clematis will also carry the largest blossoms although their overall growth is restricted. I have a *C. montana* that I cut back hard each late May, otherwise it gets under the tiles of the roof above, but the result of my efforts is that each bloom carried is 8 or 9cm across instead of the more normal 5 or 6cm on an unpruned specimen.

I think *C. armandii* is all the better for regular pruning. You can train it to cover an allotted wall space and subsequently cut all its flowered trails back to this framework immediately after flowering, in mid-May. You may be grieved to find you cannot avoid cutting its young shoots at the same time, but you shouldn't let this worry you. More will come in next to no time, and the whole plant will wear a rejuvenated appearance. After all, the old leaves on this evergreen species do look pretty tatty at winter's end.

Since writing the above I have had an interesting correspondence apropos, so I shall quote it fairly fully. Mrs Davies lives at Haddenham, near Aylesbury, Bucks, and wrote on 24 April.

'Dear Mr Lloyd, The autumn before last I bought a *Clematis armandii*. It is the success story to end all success stories but I'm now writing to ask for help. It made about 15ft of growth last season and this year has been a mass of blossom. It now looks all set to repeat its growth pattern and I am beginning (while enchanted and proud) to feel slightly hounded, as I spend a good many of my waking hours tying in and generally striving to control this abundance. I live in a bungalow and I read that this clematis should not be pruned; it is now beginning to riot over the roof! Where I should be very grateful for your help is in advising me (a) whether and (b) when I can prune it a little? I enclose s.a.e. for your reply.' I wish everyone would think of that.

I told her to prune all flowered trails hard immediately, as I have just been telling you, and asked her to let me know the result in the autumn, as the information would be useful for my book. She wrote on 25 September.

'Early in May you advised me on pruning *Clematis armandii* and asked for a progress report in the autumn. I cut it back very hard then as you suggested. It has responded well, putting out plenty of new growth. It then produced a second flush of growth in August, which showed signs of being healthy, plentiful and vigorous. At this point the saga takes a sad turn. What I take to have been grey squirrels have systematically stripped *all* this new growth – they have done the same to *C. calycina* ... They don't take the early flush of growth, only the later. Even so, *armandii* has made about 6ft in every direction (it is trained round a bay window) and looks splendid.'

I replied that earwigs, not squirrels, were certainly the culprits, that they never damage early growth because their numbers have not built up when this is made, but that they are particularly troublesome on young shoots late in the season, and had been particularly numerous and tiresome everywhere in the year in question. However, I was able to reassure her that the first flush of growth would be quite sufficient to provide a handsome display in the spring.

After that *histoire* let me return you to base. There are many circumstances where you'll never want to prune even a montana (though never is an absolute word that I hesitate to leave unqualified). If it is to romp through a tree you will just let it romp. Even so, after twenty years or more (here comes the qualification) a time may yet arrive when an old montana is no longer giving you the pleasure it once did. New growth piles up on an ever-increasing substrate of old, dead debris. At this stage the only sort of pruning that may seem at all on the cards is the complete removal of everything within sight except for 2 or 3 metres of old trunks at the base. My advice is: go ahead and do it. There is a risk of killing the plant. There is a risk of greatly weakening it so that the response in terms of young growth from very old wood is half-hearted and slow. In that case finish the job and replant with a youngster. You may, on the other hand, get satisfactory rejuvenation of the old plant, and the risk is worth taking because you'll never be happy again with an old monster once you have noticed just how monstrous it has become.

The same applies to *C. macropetala* and *C. alpina*. They don't enjoy being cut back into ten-year-old wood, but it may still be the wisest course. I recently cut an old, really thick-stemmed *C. alpina* 'Columbine' back to o·5m. It has responded but weakly, with thin spindly shoots from the old stumps. Had I made a practice of clipping it over every year after flowering, all would have been plain sailing, though tiresome for me at a busy season. As it is, the plant may or may not pull itself together. I applied the same treatment to an old *C. macropetala* (in the interests of this book, actually). From the cuts I made into 1cm-thick wood, it immediately made new green buds, but when the young shoots were about 5cm long it changed its mind and the plant died.

In sum, clematis are pretty good about breaking from old wood if they have been made accustomed from their youth to an annual massacre. But if they are left to their own devices for many years and then massacred, their response is in greater doubt. Another example: I have a friend to whom I gave a 'Lasurstern' layer many years ago, and for many years it grew and flowered well against his house without more than a little pruning and trimming. But its legs became bare and stemmy. Because of a hedge growing in front of it, it was rather dark in its nether regions and this was the probable cause of its legginess. 'Lasurstern' generally breaks well and frequently from low down, right through its life. Anyway, I recommended him to cut this old chap hard back to a foot or so (we weren't metric then), and he did. And it did break, but weakly. It was never the same again and gradually faded away. He never reproached me, either: a true sign of friendship.

With 'Lasurstern' as guide, we move on to Group B and take note of the fact that Mr Fisk, in his handsome catalogue, tells us that none of this group need pruning. In fact, doubtless appreciating that customers like to be told that a quiet life is good for them and the best way of living, he has reduced all his pruning recommendations to one of two succinct words; either 'Hard' or 'None'. Makes me wonder if this chapter is just a lot of padding.

So, according to him, it's None for 'Nelly Moser' and all her May–June-flowering companions. 'I want a large-flowered clematis that doesn't need pruning' is a request that often comes my way. In vain for me to protest that the clematis which absolutely demand hard and regular pruning are much the easiest to deal with. They won't believe me.

OK, sell them a 'Nelly Moser', tell them it wants no pruning and let them get on with it. As she flowers on the previous season's wood, she'll flower all right but in a very few years we shall be back among the birds' nests and the old mattress. Then we shall be forced into drastic action; we shall at best lose most of one year's blossom and at worst be unlucky as in the case of the 'Lasurstern' on my friend's house.

It is far preferable to do some pruning every year. If you are really conscientious about the way you do it, this is a fiddling task, so be sure to choose mild weather (the season being February or early March) and for preference a nice bit of sunshine warming your back through your winter woollies.

Remove all dead shoots or pieces of shoots. Clematis make far more growth in one year than they will ever carry forward into the next, so there is always plenty of dead to remove. Especially where they have flowered. The 'short growths from last season's wood' on which the flowers were borne will generally die at the end of the season. Long, new young vines will have taken their place and are what need to be cherished for the next season's blossom.

All the healthy young wood that is to be saved will be clinging together in a tangled skein. Snip carefully between the interlocking vines so that they can be separated and trained in various directions in order to cover the maximum possible area. This is a nerve-racking job as from time to time you'll find you've accidentally snipped through a live and healthy shoot. But you must develop the quality of phlegm. Surgeons, after all, do far worse things to their patients and it really matters, sometimes, with them, but the surgeon remains a very jolly sort of person with a schoolboy sense of humour. Surgeons make excellent growers of clematis.

The point of all this separating and training is that it enables the clematis to present its blossoms in the most flattering manner: not all bunched up and overlapping but so that they can *all* be enjoyed.

Now, I realize that one doesn't necessarily have time for this caper. In that case, having cut out all dead wood and shortened the rest to a pair of strong buds, as I instructed at the beginning of the chapter, plaster what remains against its support as best you can without further disentanglement. That'll do well enough. The main thing is that you'll have done enough to prevent the clematis from accumulating a mass of its own dead debris and thus developing into the dreaded dropsical lump.

What would happen if you made a practice of pruning Group B clematis hard back as though they belong to Group C? That is a very interesting question, as the politicians say, playing for time in the face of an awkward poser.

In some cases this treatment will provoke another perennial question: 'Why does my clematis not flower?' If it is one that insists on flowering

TOP: *C. florida* 'Alba Plena'
BOTTOM: 'John Huxtable'

TOP: 'The President'
BOTTOM: 'Marie Boisselot'

TOP: 'Pagoda'
BOTTOM: 'Victoria'

'Fireworks'

TOP: *C. viticella*
BOTTOM: 'Mrs Hope'

TOP: 'Miss Bateman'
BOTTOM: *C. orientalis* L & S 13342

'Hagley Hybrid' and *Hypericum* 'Hidcote'

TOP: *C. montana* 'Picton's Variety'
BOTTOM: *C. fargesii* var. *souliei*

Foliage of *C. recta* 'Purpurea' and Oriental poppies

on its old wood or not at all, it'll never have any old wood to flower on and hence won't flower. Q.E.D.

But the fascinating thing about clematis is that so many of them won't obey rules; they make their own as they go along. Some of our Group B lot can flower on young wood as well as off those short growths from the old (and when we say old, we merely mean shoots that were young in the previous year: year-old wood, in fact). 'Nelly' and 'Marcel Moser', for instance. Even if they've flowered freely in May–June, healthy specimens can be expected to carry at least one bloom (sometimes several) at the ends of the current season's new vines any time from August to November. A hard pruning of all the previous year's growth, in February, will pre-empt the entire May–June crop but will tend to swell the numbers carried in late summer and autumn on the superabundance of young shoots. This is quite logical. The May–June flowering is admittedly what we are generally aiming for in Group B, but the production of a mass of blossom then has a very inhibiting effect on the subsequent production of young shoots. I know of a 'Miss Bateman', growing in a rather tight, starved corner, that flowers so heavily in some years as to be incapable of producing any young shoots for the remainder of that year. Consequently it hardly flowers at all in the next year.

So this coin has two sides. If the 'Miss Bateman' I write of were sufficiently well nourished and watered, it would be able to flower and shoot well all in the same season and hence every season, and that would be ideal in most cases.

However, supposing you were taking your holiday in the South of France in May–June. What use would it be to you to be told that your 'Nelly Moser' had never flowered so well as in your absence? To rub salt into the wound, 'they' might take a coloured photograph of it to show you what you had missed and to impress on you the folly of being out of England in May–June and especially when, crowning insult, the weather on the Riviera turned out to be far worse than it was at home. Well then, you could put a stop to at least one opportunity for *Schadenfreude* by pruning your 'Nelly Moser' hard back in February. There'd be no flowers in your absence and perhaps an increased bonus crop in the autumn. 'Lasurstern' is even more strongly predisposed to flowering well on its young wood in August–September if pruned hard in late winter. I know of two instances near me (ones I drive past) where this is its regular treatment. They grow next to the front door and I expect their owners dislike their winter appearance and cut them back in self-defence. They're missing a lot but their method undeniably works: the late summer display is good.

And so we edge towards those varieties which the pundits will instruct you to cut hard back as a regular routine but which I should be inclined to give a double chance. A keen clematis-growing friend in Dorset wrote me one February: 'Have you read what Tony Venison says about pruning clematis in today's *Country Life*? I never knew that 'Mrs Cholmondeley' was jackmanii group and it is always the first of the large ones to flower with me. I shall most certainly *never* cut it, or rather her, down.' If your 'Mrs Cholmondeley' is well fed and watered, she will flower abundantly both early and late, given the Group B treatment, but there may be circumstances when it is preferable to treat her as Group C, thus limiting yourself to the one crop.

The same with 'Ville de Lyon', a variety that puts out distress signals sooner than any other if at all deprived of food and drink. All the lower stem leaves die prematurely (and remain on the plant, of course) which can look most unsightly if it is grown on a wall. If you cannot avoid this condition (but you can), then do as the pundits tell you. Cut 'Ville de Lyon' hard back annually and let it concentrate its efforts on just the one, late crop. But if it is growing in a well-manured border and luxuriating with an abundance of growth, a light pruning will enable you to enjoy early blooms and these are much the largest and hand-somest of the whole season. Neither will there be any shortage of late blossom. It will continue through most of the summer and early autumn.

I draw attention to this potential in other varieties where I deal with them individually. There is one further point to emphasize here. Nearly all the large-flowered double clematis carry their double flowers on short growth off the previous season's wood. If you prune them hard you'll get no double flowers, only singles ('Duchess of Edinburgh' and 'Miriam Markham' are the only exceptions I can think of, and the small-flowered doubles of viticella affinity do come double on young wood). This would be disastrous in varieties like 'Vyvyan Pennell' and 'Proteus', where the main charm or excitement, rather, of the flower is in its doubleness. It would matter less in 'Daniel Deronda', 'Jackmanii Rubra' or 'Jackmanii Alba', where the early double flowers are a bonus and the main crop is single, however lightly you prune.

We can now, at a stately pace, move forward to the late flowerers, those which flower wholeheartedly on the last few metres of the young growth they have made in the current season. By pruning them hard annually you not merely clean the plant up but you encourage it to put on masses of productive young growth. Many amateur gardeners object to this pruning. In the first place, they don't realize how easy and quick it is. You simply make a series of vicious cuts at a low level, then grab the severed stems in both hands, pull downwards and outwards and away they all come from their support, leaving only a few random leaves behind to tell the tale. Even if their support is a climbing rose or an evergreen ceanothus, say, the clematis will break away from it cleanly and sweetly.

The amateur's second objection is that, as he has expressly planted the clematis to cover a piece of blank wall, he doesn't want to have to expose it again, which is what this type of pruning entails. What he fails to realize and what the nurseryman/salesman can hardly be expected to explain, is that, at the time pruning should be done and for months afterwards, the wall will look better without a shabby clematis on it than with. If the clematis belongs to Groups A or B, he'll have to endure its prolonged dishevelled appearance in the winter half of the year willy-nilly, but if it is a Group C clematis he can give himself a blessed rest from it.

But he doesn't like his bare wall? Then he shouldn't grow a clematis on it – not in isolated squalor, which is what I've been trying to say all along. First cover the wall with suitable clothing, for goodness sake: cotoneasters, ceanothus, chaenomeles, climbing roses, figs, wisterias, magnolias, myrtles; their name is legion – and then grow your clematis over them.

The clematis in Group C lend themselves particularly well to the pole treatment: as vertical features, a column of blossom rising above lower surroundings. The clematis is cut annually to within 0·3m of the ground. As it makes its young growth in spring and early summer you tie this in to the pole (you'll need a step-ladder for the top ties because the pole can rise to 3m above ground level). Use a heavy grade of tarred string for these ties. If the string is soft and fluffy (as is fillis) sparrows will tear it off and use it as nesting material.

If the clematis is to be grown into a small tree, you may not want to prune it so low, as extra height is required of it. Always prune low in its first year, but in subsequent years you can ease up a bit and leave a greater length of old stem each time. But there are few occasions when this is necessary.

Although *C. orientalis* and *C. tangutica* belong to Group C there is quite a lot to be said for pruning them fairly lightly, back to a permanent framework that may reach up to 2m. This is because of their non-stop habit of flowering once they have started. Light pruning enables them to start before June is out, and they will then continue right into October or even later.

Most of this group have one long but limited flowering season. It is longer than the comparable season of *C. montana* (Group A) or the first flowering of 'Nelly Moser' (Group B), because in those the buds all form at one and the same time and all open more or less together, whereas in Group C we have panicles of blossom in which the centre ones of each cluster open first and are followed at a distance by subtending buds and these, in some cases, by buds subtending the subtenders. But when all these have flowered there is nothing more to follow, and the revels are at an end even though autumn may scarcely have arrived.

If you are growing a Group C clematis against a wall, you can greatly increase the area it will cover by intelligent pruning. If you cut all the vines back to the same point, the young growths will all rush up together in one vertical skein – just what you want when a pole is the support but seldom what is required on a wall, unless there are windows on either side. The way to prune in this case is to cut the shoots back by different amounts: some hard, some less hard and some so as to leave as much as 2m of old stem. Then they are trained fan-wise to right and to left, leaving the shortest in the centre and keeping all as low as you conveniently can. The young shoots will, as always, develop from the extreme pair of buds above which each pruning cut was made and will run up the wall like the prongs of a fork, covering a large area.

I am often asked what can be done to prevent a clematis from rushing up a wall and flowering out of sight at the top. This habit is usually most noticeable on dark, sunless walls. The clematis is rushing upwards in search of light. Group C clematis are the most likely to frustrate you by this disappointing performance. There is little you can do to control their growth while they are actually growing, and they flower at the height that pleases them while they are still growing. Group B clematis can be more easily persuaded to flower where you want them to. Even if their new growth has run up too high for your sight lines, you can re-distribute and train it against the lower wall face at pruning time. The flowers will be borne on short growths from this re-distributed old wood and you'll have them where you want them.

There are refinements worth discussing on the season of pruning. Apart from Group A, whose pruning is delayed until after flowering, all clematis are normally pruned in February or March. By then you can see what's live and what's dead because the live shoots will be on the move. Most clematis are very precocious in this respect. It is particularly important to know what's living and what isn't in the pruning and training of Group B, so there is little scope for choice of other pruning months between autumn and spring in their case. You could do it in January but ... I am writing these words in January – a very mild one – and the Group B clematis are already amazingly forward. But if I sorted them over now I should inevitably expose their young buds to a greater extent than they are exposed before pruning, when the tangle of old, dead shoots does afford them some protection against frost. Every one of these green Group B leaf clusters that are now plain to see contains a flower bud: not yet visible but there just the same. If I lose it through frost damage in February or March, the loss is total and without redress, so I shall walk delicately and delay my pruning yet awhile.

Not so with Group C clematis. These are the most hideous of all in their off season, a tangle of dark shoots and blackened foliage that they refuse to shed. To endure this spectacle throughout the winter is asking a lot and I seldom do. I cut them hard back – give them the full pruning treatment, in fact – at any time that suits me from when they become dormant, in November, onwards. This will do no harm to a strongly established specimen. In fact it is much like pruning your roses early, as is nowadays widely practised, instead of leaving them to the traditional March season.

It is true that, having been pruned early, Group C will sprout new young shoots dangerously early from just below where you made your cuts. I have just, late January, checked on this in my rose garden where I pruned the clematis at the same time as the roses a month ago. 'Perle d'Azur' and 'Star of India' both have young shoots on them from 15 to 30cm long. 'Mme Julia Correvon' only 13cm long and less still – 3cm – on 'Kermesina'. The most forward are sure to get frosted or battered by wind before we're out of the wood, but this won't matter. These are shoot buds, not flower buds. If they are destroyed, others behind them will develop and take their place. I shall have lost nothing.

I have said that most Group C clematis flower themselves to a standstill, after which nothing more can be expected of them, even though their flowering may be at an end by, say, mid-August. Southern gardeners of an experimental temperament might like to try for a second crop by cutting everything back by half as soon as the first (and generally only) flowering has drawn to a close. If the autumn remains open and warm as autumns so often have, of late years, the clematis will sprout again and have time for a second blossoming before the growing season ends. At least, this worked for me on a *C. viticella* that I so treated, and it should be feasible on the earlier Group C performers like 'Etoile Violette', 'Hagley Hybrid' and 'Comtesse de Bouchaud'.

Another refinement. You were told by Christopher Lloyd that 'Etoile Violette' and *C. flammula* associate prettily together, the froth of white of the one offsetting the substantial violet flowers of the other. On trying this out, however, you found that 'Etoile Violette' flowered so much earlier than flammula that there was no overlap. So you write to C.L., acidly pointing out the error of his recommendation (this actually happened), and he, being pleased rather than distressed at having peeved you, does not reply. But had he liked your tone he could have redressed matters by first apologizing and admitting that 'Gipsy Queen' would have been a far more suitable team-mate for *C. flammula* than 'Etoile Violette', flowering at least a month later. But, as you have already got the first suggested pair growing merrily together, you can manipulate their behaviour. Prune flammula only enough to remove its last year's old flowering panicles, thus encouraging it to flower earlier than normal. Contrariwise, prune 'Etoile Violette' extra hard and extra late, say in April, thereby very considerably delaying the onset of its flowering season. By dint of two opposite pruning treatments Cox shall meet Box at last.

Incidentally, whereas it is so often a good plan to plant two clematis of the same pruning group side by side, it is not so clever to grow a

Group C clematis near to one belonging to Group A. I have done just that: *C. viticella* and *C. cirrhosa* var. *balearica*, cheek by jowl. When I prune *C. viticella* in the winter, it is quite easy to pull its shoots out and away from *C. cirrhosa*, but when I want to reduce the latter, following its flowering, which doesn't end till late April, the young shoots of *C. viticella* are already threaded through it and I don't like to spoil them (I could all the same, and I shall one of these days).

There may be occasions when you want to prune Group B clematis at an unconventional season. You remember the example I cited of a 'Miss Bateman' growing under rather starved and parched conditions at the edge of a meadow. She flowers so abundantly in alternate years that she is unable to find the strength to make new shoots on which to flower in the intervening years. Apart from the obvious cure by organizing a better diet, you can redress matters to a certain extent in the pruning department by cutting all the flowered wood back immediately following her flowering. This will be some time in late June, and will encourage the production of young shoots rather than the channelling of her energies into ripening seed. Similarly, if you have failed to prune a Group B clematis for several years and it has got out of hand: instead of resigning yourself to drastic treatment in February–March and no flowers in May–June, you can wait till after the May–June flowering and hack back then, including, willy-nilly, all the young shoots that have been made that spring.

I must, before leaving this subject, bring up a question to which I don't know the answer. Some clematis, notably 'Ernest Markham', 'Huldine' and 'Lady Betty Balfour' – in some gardens or in some positions in the same garden but not in others – make masses of growth each year but carry few or no blooms. It is certainly wrong in principle to grow these in a shady position, but don't now write and tell me that you have a 'Huldine' in absolute shade that never fails to reward you with cascades of blossom. Such things can be, and shade in the southeast of England is very much akin to sun in the north. I know of examples of *C. orientalis* in Ludlow and Sherriff's form (13342) that flower excellently in shade in Sussex and Kent but do nothing more than vegetate in the north, especially when shaded.

In the case of a variety like 'Lady Betty Balfour', season makes all the difference. I should never dream of planting this clematis north of Trent anyway (but see p. 49), but even in the south and in sunshine she is apt to miss out following a dull summer, but makes a fantastic display after a sunny one.

Sunshine comes into it, then, but is not the whole story and it is frequently suggested that pinching out of young shoots when they are 2m long will snap them into a flowering condition. Particularly is this suggested of 'Huldine' and 'Ernest Markham'. I can only say that such has not been my experience. If one of these clematis is a recalcitrant non-flowerer in one part of your garden, however, do move it and try it in another. As with the temperamental *Gentiana acaulis*, this may just do the trick and if it does, you won't scratch your head too much in search of a reason. Results speak for themselves.

Young Stock

Finally, the pruning of young, newly acquired clematis. It is standard practice to shorten these back to a pair of buds (which are very likely dormant but will not remain so for even a week after pruning) 15 to 30cm above ground level, in February–March. This is a sound recommendation in most cases. It helps to build up a stout, slug-and-disease-proof stem, in the first place, and it helps build up a strong root system, preventing an overload on its resources. However, I should make an exception in the case of strong plants within Group A: *C. montana, C. alpina, C. macropetala, C. armandii,* etc. If these are well grown they should be able to cope with a metre of stem and there'll be flowers to enjoy in the spring.

Should you buy your clematis in spring, they will already have been pruned on the nursery and there will be no need for further shortening back, provided the roots look strong.

Chapter 9
PROPAGATION

'How do you propagate your clematis? It's quite simple, isn't it?'
Thus the visiting public to Lloyd, hard-pressed and a trifle
brusque, on various occasions between spring and autumn. When you
are at the seat of custom and there are others waiting their turn, it is
impossible to stop in mid-stream (the current uncomfortably strong)
in order to discuss a subject that will occupy quite a long chapter here.
By refusing to discuss it, I am not demonstrating a cagey secretiveness
in respect of methods that I intend to keep under my hat. Far from it.
I immediately refer my questioners to my book and suggest they should
borrow a copy from the public library. If they are interested enough in
the subject, this will not be too great an effort. If they are not all that
interested, neither need I be in them.

In the first book to be written on clematis by Moore and Jackman
about a century ago, propagation was not so much as mentioned, let
alone discussed or described. That a nurseryman should reveal his
methods was inconceivable in those days. Indeed, this particular iceberg
has only melted in the forty years since the last war. But it has melted,
if not completely, and this says a great deal for human nature. It is the
kind of uncelebrated progress that will never make news but that I
find marvellously heartening. There is a flourishing and expanding
organization called the International Plant Propagators' Society, at
whose meetings nurserymen and representatives from research insti-
tutes and teaching colleges exchange details of propagation methods
with complete lack of reservation and without thought of rivalry. It is
true that the press are not invited to these meetings, but the proceedings
are published in due course and from then on there is no objection to
anyone anywhere being informed of what took place.

I find it hard to explain the reason for this changed approach. The
younger generation is free of jealousies that were, until very recently,
commonplace. They appear to realize that if you are good at your job
you have nothing to fear from others knowing how you do it; that

shared knowledge both among professionals and with the consumer public is stimulating. I think nurserymen as a body are probably nicer people than they used to be. Whether they are nicer because they have become more communicative or whether being nicer has made them more communicative, I don't know, but I can say that as a body, the trade and all those connected professionally with horticulture are lively and pleasing company. I feel sure that two reasons for this are, first, that they know they are doing a worthwhile and interesting job; second, that it is an active industry wherein methods are changing and improving at an incredible rate. Clematis production has not lagged in this respect and the plant's popularity has been met by a greater efficiency in its propagation than could have been dreamed of forty years ago.

I will start by discussing the subject at the professional level of mass production, and then consider what methods are open and most easily accessible to the amateur.

Grafting

Until thirty-five years ago, the principal method of clematis propagation was by grafting. Only very easily struck groups like the montanas, alpinas and macropetalas were rooted from cuttings.

The most commonly used stock for grafting is *C. vitalba*, with *C. viticella* as a popular runner-up. Seedlings are imported from Holland. They are seldom raised in this country, but from direct sowings in an outside seed bed are ready in their second year, being ideal for use when 5 to 8mm thick where the graft is made. January to March is the main season, and the stock plants for producing scion wood are brought into early growth under heated structures. Their young trails are ready for use when about a metre long, and can then be expected to yield ten scions (or more) each from five reasonably firm nodes, the tips being discarded. The nodes are split longitudinally and cut with a shield of wood behind each bud. The stock is prepared by cutting across the stem so as to leave 3cm of collar above the roots and a simple whip graft is made. It is tied in from the bottom using wet raffia or, if this rots too quickly, carpet thread is suitable (obtainable from a haberdashery). The grafted seedlings are potted into very small pots, twisting the roots so as to get them in and leaving the scion bud just above the soil surface, where it is less likely to rot than when buried. The pots are stood on a bench (or floor) with bottom heat of 21 degrees C and covered with clear polythene, this being turned daily so as to obviate undue condensation and, once again, the probable rotting of the eye.

When young growth starts in about a month's time, the temperature is lowered to 15 degrees C and hardening-off is gradually achieved. At eight weeks the young plants can be given their final potting into 10cm long toms and they will make strong plants for sale in the autumn. It will thus be seen that the whole operation from grafting to the production of a saleable plant is quick: nine months or less. What may be another advantage is that grafting can be done at a comparatively slack season on the nursery.

The method has been widely criticized on the principal counts that grafted plants are harder to get established in the garden than 'plants on their own roots', that they are more liable to contract wilt disease, that collapse of the scion is seldom followed by regeneration from below ground level and, in sum, that so great is mortality of plants raised in this way that many have given up clematis cultivation in despair. Thus, Ernest Markham (*Clematis*, pp. 31–2). I agree that many have given up the cultivation of clematis in despair, but this has nothing whatever to do with the practice of grafting. The Glasshouse Crops Research Institute's experiments (see their Annual Report 1965, pp. 128–31) on the susceptibility of clematis to infection by the fungus causing wilt disease revealed no greater inroads by the disease on grafted plants than on plants raised from cuttings.

Indeed, why should they? You graft on to a seedling whose roots provide the initial boost that enables the scion to get started. The graft union (below the scion bud) is buried beneath the soil surface and the scion quickly begins to make its own roots. In a short time these greatly exceed the stock's roots, and by the time sales take place the customer is buying a plant that is virtually (80 per cent, anyway) on its own roots. The graft was a nurse graft. The stock does not provide a root system for the plant's whole life (as is generally the case with roses, for instance), but merely for long enough to enable the scion to take over, root and branch. When planted in the garden, a grafted plant should be set low enough for 3cm or so of stem above the highest roots to be buried below the soil surface, and this is a principal safeguard against damage (dogs gnawing) and disease (usually fungal) on the aerial portions. If these collapse or disappear, dormant shoot buds underground will break and off you go again.

These dormant buds, let it be emphasized, will, even on grafted plants, belong to the scion, not to the stock. They will not be undesirable suckers. 'When a grafted plant collapses this is usually its end, and not infrequently the growth of the stock appears in its place.' Markham, again. Now grafting is done on that piece of seedling stem between its roots and the highest pair of leaves (seed leaves, known as cotyledons) called the hypocotyl. It is impossible for the stock of a clematis to sucker from here or from its roots. This could only happen if the graft were made high up the stem, above the cotyledons. No doubt that did happen, occasionally, and it is an example of grafting badly practised, but no indictment of the practice of grafting. I have not met a case of a grafted clematis throwing suckers in the past thirty years. (We must not allow ourselves to be confused by the ability of *C. macropetala* and *C. alpina* to sucker on their own roots. These species are never used as stocks for grafting anyway.)

Where grafting still persists as a commercial method of clematis propagation it is for one of two reasons. If a nursery has always been geared to this production method and finds it satisfactory, there will be an inbuilt disinclination to change over to another method even if there is a chance or probability that the latter will prove more effective. A considerable disruption in organization will be entailed, expensive change-over in equipment and a re-deployment of staff duties. This kind of disruption is not lightly embraced.

The other reason for the survival of grafting as a method for raising clematis is that the provision of a ready-made seedling root system enables a great surge of energy to be unleashed on the scion bud. If the scion is habitually weak-growing, as is *C. florida* 'Sieboldiana', or if there is difficulty in making the scion bud break dormancy, as in *C. armandii*, then this extra boost may make all the difference between success and failure.

C. armandii being a special case, I will consider the problems it presents to the nurseryman at this point. It is a clematis that he cannot ignore. The public regards it as a highly desirable evergreen, the only clematis combining evergreenery with vigour, reasonable hardiness and a good floral display, so it is astonishingly popular (I am astonished, anyway) and demand unfailingly exceeds supply.

When raised from cuttings, *C. armandii* roots with little difficulty, but it can be the devil's own business to make it shoot. I have known cases where cuttings have taken so well that the roots almost fill the pot but, come the winter, the leaf dies, shortly followed by the node and stem below. All that remains is a bundle of strong, healthy roots – infuriating! Most armandiis *are* raised for sale this way, but results are uneven and depend to a large extent on the clone used.

The shortfall can be made up from seedlings. In 1986 a Hertfordshire firm was advertising *C. armandii* seedlings at 60p each. In such a variable species this is a dangerous practice; very small and miserable-looking flowers are likely to result.

Some nurserymen find grafting an excellent method. In this case the stock plant (or plants, but with such vigour, one could well be enough) is generally grown outside on a wall. Its young growth is too coarse and vigorous to be used just like that, so, when about six weeks old, which would be in June (I am citing an actual example) all young trails are shortened by half. The resulting growths are of just the right thickness and they are less likely to carry flower buds. Leaves with axillary flower buds (recognizable by their plumpness) should always be rejected. Each trail or vine should yield fifteen to twenty grafts. Grafting can be done from December onwards in the manner already described. I have bought in some of these grafted plants myself and they are of excellent quality. It doesn't matter if they are rather shorter at planting out time than most other clematis. As long as an active shoot has been made, it will romp away on finding itself with a free root-run in the garden.

Cuttings

All clematis can be propagated from cuttings, and in retrospect it seems odd that this basically simple method should only have gained widespread currency in the last few decades. It is not as though we had to await the discovery and introduction of more sophisticated equipment, even. Propagation under mist is now generally accepted as being less successful, in the case of clematis, than by more traditional and obvious methods. There is too great a tendency for foliage to damp off under mist, especially where there is any leaf overlap. Protective fungicides anyway need to be used regularly from the outset, and I suppose the greater efficacy of modern fungicides, compared with what were once available, is the one (not very dramatic) tool in our modern armoury that has contributed to a high measure of success in the progagation of clematis from cuttings today.

The method used by professionals is basically as follows.

Propagation material for cuttings is taken from plants raised the previous year. This is general practice nowadays, and not just with clematis. Young material from young plants gives much the best take, far better than is obtained from old stock plants kept specially for the purpose. Old plants take up a lot of permanent space and their growth is far coarser, the leaves inconveniently larger than in young plants. The latter, after serving as propagating stock, can simply be grown on for sale in the usual way, so there is a complete and constant turn-over of stock (provided you can sell it of course).

The young stock is kept in greenhouses which need not be heated. Even in the north of England, stock plants are ready to yield plenty of cuttings material by the middle of April. Growth which began in earnest in late February resists the frost that will enter greenhouses from then on.

The propagation bed is set on or into the floor of a greenhouse or polytunnel, electric heating cables being buried 4cm deep in sharp sand which is sometimes topped off with a finer-grade horticultural sand to allow better thermal contact between the bed and trays. Thermo-statically controlled, a consistent bottom heat of 24 degrees C is applied when the cuttings first go on to the bed in April or May. The tem-perature is progressively reduced after rooting to end at 16 degrees C by late June, just prior to potting up.

The cuttings are internodal (I will explain about this in a moment) as is normal in this country, and consist of one pair of buds, one leaf (the other is removed) and up to 5cm of internode (the stem of the cutting). If the latter is hard, it is wounded, but not otherwise. Wound-ing consists of the removal with a knife or other blade of a shaving

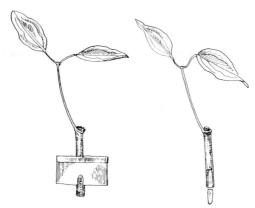

Wounding. To help root production, if the stem of the cutting is hard, slice off a shaving about 2 or 3 cm long at the bottom of the internode. *Right*, a sliver removed.

of rind and soft underlying tissue from the side of the bottom 2 or 3cm of the internode. Exposure of this larger area facilitates root production.

If the leaf that remains on the cutting is rather large, the central leaflet can be removed or reduced by half. That this may be necessary will be understood when I go on to say that 100 cuttings per standard seed tray is average, and one wants no overlapping of leaves (otherwise the bottom leaf doesn't get its fair share of light and cannot function properly).

Wooden seed trays (boxes or, as they say in America, flats) are used, the bottom covered with 6mm of gravel, then the cutting compost, which could consist of two parts coarse sand and one part peat, but every nursery has its own slightly varying recipe and some include a bit of soil. I find the old John Innes recipe satisfactory: one part by bulk of sterilized loam, two parts peat, three parts sand.

The cutting compost is pressed with a board and then topped with horticultural sand, this being levelled with the hand. Sand, being inert, helps to prevent the development of mosses and disease while the cuttings are rooting. The newly made cuttings are put in a colander and dipped in a Captan or Benlate suspension. The base of each is then dipped in a rooting powder. I have used Seradix B2, and the multipurpose types, made by Murphy and ICI, seem just as good. To be effective, they must be fresh; after two years their potency can no longer be depended upon. I suggested in an earlier book that manufacturers should print a 'use by' date stamp on their product but nothing has come of it. I would have thought it in *their* interests too; they would probably sell more in the long run. The cuttings are pushed into the compost. If so soft that they cannot stand this strain without bending, they are too soft to be of any use anyway. The bud should just rest on the surface of the rooting medium. Cuttings are watered in with a suspension of Captan, Benlate or Rovral to prevent fungal infection. They are damped over periodically thereafter from a watering can to minimize transpiration and prevent wilting. How often depends on the weather; it might be once a day or as often as eight times on a roasting day. The professional and skilled amateur develop a 'feel' for this, adapting the aftercare of their cuttings to the sunshine and humidity prevailing at the time. Clear polythene sheets, either on frames or in the form of cloches, are initially drawn over the cuttings to keep them close and this helps to reduce the need for damping-down.

Shading is just as important. It needs to be effective in preventing sun scorch but, short of this, the more light cuttings receive the quicker they will root. Some nurserymen quite simply shade the glass through which direct sunlight might enter with one of the proprietary white liquids, applied with a brush or spray. Others use overhead hessian screens.

With any soil-warming systems care has to be taken over watering. The cuttings dry out from the *bottom up*, and compost which looks wet on the top can be bone-dry 6cm down. A thorough watering is given every few days and includes a fungicide to prevent botrytis. Since resistant strains of this disease readily develop, the fungicides Captan, Benlate and Rovral are used in rotation.

The first week is critical; if the cuttings look well at the end of seven days they will root eventually. However, if a prophylactic has not been used, botrytis will more often than not take hold underneath the leaf canopy. It will only become visible after it has destroyed much of what you cannot see. If the cuttings are set so that there is no overlapping foliage, and all receive an equal share of light and air, the problem is greatly reduced.

Within ten days a cutting removed from the bed will show callusing, which is a necessary precursor to rooting. This may begin only a few days later with the faster-rooting cultivars.

After four weeks, roots begin to appear through drainage holes in the tray. Hardening-off takes another two weeks or so, and then the cuttings are either potted up (John Innes No. 2) into 10cm pots or left *in situ*. The latter will be lifted the following February, separated, washed in a fungicide suspension and potted using the same John Innes No. 2 compost. The former develop faster. Many small-flowered species and varieties are on sale by late summer.

There is a continuing tendency for nurseries having no ready supply of loam to use a peat-and-sand potting compost with all nutrients added. I have seen some very starved-looking clematis grown in this way, but methods are all the time improving and if slow-release fertilizers are used and feeding is not neglected, as necessary, excellent results can be obtained.

Internodal cuttings, as I have already remarked, are the type used in this country, but in America (the US anyway) nodal cuttings are often preferred. I must explain. A nodal cutting consists of two nodes (or joints), each with a pair of opposite buds and leaves. All the leaves except one of the upper pair are removed. The lower node is buried in the rooting medium at such a depth that the upper is just resting on the surface of the cutting compost. However, if the distance between nodes, i.e. the internode, is long, the method is inconvenient, since the top node will be waving about in mid-air. Fortunately clematis are as

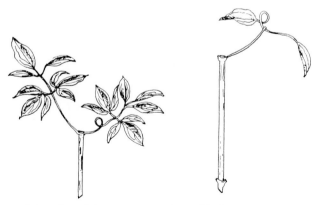

Left, an internodal cutting of the sort commonly used in Britain; *right*, the nodal cutting preferred in America.

ready to root from a cut made at any point along the internode as they are from a nodal cut. So the internode is used as an anchor and is tailored to be of any length that suits the propagator – commonly 4 or 5cm. It will also be appreciated that by using only one node to each cutting, twice the number of cuttings can be made from any given quantity of propagating material, as can from nodal cuttings employing two nodes. If you are short of material you can double up on it yet again by slicing each internodal cutting longitudinally down its centre with a thin blade. Each cutting will now consist of one leaf, one bud and a length of bisected internode. You will not do this unless you urgently need to increase stock rapidly from a small amount of material, because the fact that each cutting made comprises only one bud instead of two reduces its chances of survival.

And by the same token the American propagators prefer a nodal cutting having two pairs of buds because, under their severer climatic conditions (hotter in summer, colder in winter) the upper pair of buds often dies but the cutting is able to regenerate and shoot from the lower, buried buds. The American nurseryman will propagate from very newly rooted material at that stage when it is still packed 100 to the seed tray and before it is potted off. The young growth at this stage has small leaves and short internodes, so there is no problem in handling the material except the business of separating the vines, which will all be clinging one to another. The nurseryman I spoke to assured me that they found this no problem, and that his women separated the vines quickly and efficiently and without damaging their foliage.

Micropropagation

I give a brief mention of this as a method that we must hope will gain more currency in the future. A few firms are propagating the more difficult species (*C. florida* 'Sieboldiana' is one such) by micro-propagation but there seem to be difficulties with the entire *Ranunculaceae* family, not just with clematis. Richard Pennell recently wrote Tom that, despite many promises, several firms specializing in the work have failed to deliver.

The technique belongs more to the laboratory than the greenhouse, and involves treating plant cells from shoot tips with hormones to induce division under sterile conditions in flasks. Kits for amateurs do exist, but I am told that they are expensive and not infallible.

If successful, micropropagation might be a good way to rid clematis stock of virus infection as is widely practised with fruit. The growing tip of a stem is virus-free; cells from this region, when cultured, will produce clean plants. It is suspected that a number of clematis have been weakened by viruses (see also p. 155). It would be interesting to see if they would be improved following micropropagation. However, the method is so expensive where only small quantities of any one cultivar are required, that it seems unlikely, at present, to gain widespread currency.

The Amateur Propagator

Where does or how can the amateur get a look in? He has no stock plants except the one growing in his or a friend's garden. I have said that the material from this will be unsuitably coarse and anyway he probably has no bottom heat, perhaps no frames, no polythene tunnels. Layering will be his best recourse if only the odd extra plant or two is needed, but it is not always convenient to make layers in a friend's garden and return for them when they have rooted. I have done it, but the friend needs to be a close one.

Let us look into the possibilities of cuttings without tears first.

If you have no propagating equipment, then rather than mess about with a pot covered with a plastic bag I strongly recommend buying yourself several small propagator units. They are cheap, effective and couldn't be simpler in design. George Ward makes one that consists of a standard plastic seed tray with a clear polystyrene cover that fits snugly over it. The cover is a straight-sided lid that allows plenty of head-room for various types of cuttings and it is fitted, in its covering roof, with adjustable ventilators, so that hardening-off is made easy.

Bottom heat speeds up propagation but it is by no means necessary. If you are taking your cuttings in May–June (the best time) when the sun is very hot, the propagator will certainly need pretty heavy shading for a start, but once you see some of your cuttings making new shoots you can admit a little air, and the more air you can admit without the cuttings wilting, the more light you can allow them to have.

Ideally, soft and half-ripe cuttings taken in summer will benefit from all the light available that they can take without wilting and scorching. They need light in order to root, and if the humidity of their environment is high enough they can take a good deal and still remain turgid.

High humidity is also ideal for the propagation of fungus spores,

An internodal cutting with the terminal leaflet cut off to reduce excess leaf area.

and fungal disease is enemy number one in the early stages. So you must spray with a fungicide. Vary the chemical so that different strains of botrytis cannot become resistant to any one. A weekly application will be appropriate.

If cuttings material is taken from an established plant in the garden, drastic reduction of foliage area will be necessary otherwise the cutting cannot cope with such a large preponderance of leaf and will certainly collapse. So, even though we are talking about only one leaf to each cutting, this will probably need reducing by half or two-thirds. Say there were three leaflets, you could probably shorten them (a thin blade applied vertically to a hard surface is the easiest method) all by half, or you could remove one completely (especially if, as often happens, it is badly placed and tucked under the others, because of the way the petiole was twisted round its support) and shorten the other two. When you take a young clematis trail from the garden you will often find that one of the leaves at each node is clinging by its twisted petiole to a support while the other is unattached and hence undistorted. When you prepare your cutting, remove the awkward, twisted leaf and petiole and leave the straight one. Because, you see, when you have inserted your cutting you don't want any part of the leaf to lie against the compost surface, where it will be permanently wet and inclined to rot; neither do you want any one cutting or part of a cutting to shade another. Both these

The easiest way to make clematis cuttings is to place the stem on a hard surface and apply a razor-sharp blade vertically.

conditions are met if the leaf and petiole are respectively flat and untwisted.

As I hinted at the start, the most easily rooted clematis are the montanas, alpinas and macropetalas. If you insert these in May, and pot them off into 10cm or larger pots when well rooted, you will have very strong plants by the autumn, a metre or more tall, and they will flower in the following spring.

However, if your cuttings are taken rather late and are, furthermore, of large-flowered hybrids that are slower to root, they will, under cold conditions, very probably not root until the end of August or later. Never be tempted to disturb these late rooters, or you will encounter severe winter losses. Leave them in their propagating pots or boxes until the following March and pot them off individually then. This advice applies to most shrubby plants. If they're not well rooted before the end of August, let well alone till the following spring.

I don't see why you shouldn't succeed, in a small way including plenty of failures, with hardwood clematis cuttings, involving no equipment at all. Suppose you were pruning a clematis in October (early autumn is the best moment for taking hardwood cuttings of anything) you could make cuttings of its strong young wood that each consisted of a node supported by 10 or 15cm of internode. Stick them into a plot of light soil in a reasonably shady position (such as the north side of a wall) so that the node and its buds are just visible at the soil surface. Water at droughty times next year and lift in the autumn. I should use a strong rooting compound here: Seradix B3 would be suitable.

Using a different method, some varieties are quite easily raised from hardwood cuttings. The montanas, *C. chrysocoma* and *C.* × *jouiniana* respond very well.

Unusually, late winter seems the optimum time. The cuttings are internodal and about 10cm long. It pays to wound the stem; remove a sliver of rind from the bottom 3cm or so. Insert as for softwoods into your normal cutting compost but use the strong rooting compound recommended earlier. The cuttings will benefit from the protection from excess moisture that a greenhouse or cold frame will afford them but will withstand frost, so no artificial heating is needed.

By late March, the sun can easily heat the compost surface sufficiently to scorch the buds, which by now will be breaking, so shading becomes necessary. The cuttings are best not covered with polythene – it induces premature bud-break and encourages botrytis. Apply fungicides as you water. By late April/early May, roots should appear through the drainage holes of your container. Wait a week or two and then pot into $3\frac{1}{2}$-in pots using a John Innes No. 2 compost or equivalent. They are often strong enough to plant out from midsummer onwards.

This method has worked, in Tom's experience, only with the varieties mentioned above. The large-flowered hybrids and other species responded patchily. You might, however, like to experiment with these, varying the times and technique. Let us hear about *your* successes.

I have even rooted hardwood prunings in a glass of water. The variety was 'Huldine' and the time December. They stood on a windowsill in our dining-room and I changed the water occasionally when it turned green, otherwise topping it up. By June, half of them had made roots. More recently I have done the same with 'Kermesina', only in this case I stood the jam jar in a cold frame among pot plants. When the plants were watered the level of the water in the jar was automatically brought up to the rim. There were eight cuttings and five of them rooted. I

A cutting with well-developed roots.

potted them individually into John Innes No. 2 on 17 July. These cuttings were internodal and were 15cm long. If the cuttings are attacked by greenfly or other pests, rather than spray them in the house (some pesticides smell disagreeable) you can effectively wash them under a cold tap.

For any gardener requiring just the odd extra plant or two, layering offers by far the easiest and in every way most satisfactory method, quickly giving rise to a strong young specimen. Bury any live piece of clematis stem in moist soil, and it will make roots and shoots. But the piece of stem in question must be firm and mature, not soft and green; current season's wood is seldom suitable, but if you do your layering in winter (which is always a good time because there are no leaves to get in your way and you can see what you are doing) then practically any live wood will make suitable layering material. Even ten-year-old trunks may be layered successfully, though this is not the most convenient material.

You don't need to be at all scientific about layering clematis. Scoop out a hollow, say 10cm deep, in the ground at a convenient point, dip a clematis stem (not a cut-off end of stem but a middle portion that goes into the ground and comes out again) into the hollow, hold it in position by pushing in a wooden peg cut from any branch, into the ground at the bottom of the hollow, return the soil and ram home. It is wise to mark the spot with an upright stick or cane so that the layer doesn't subsequently get dug out before it has rooted.

By the following autumn you will be able to cut the layer lead – that

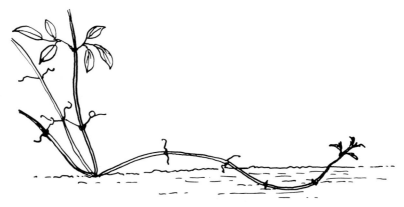

Layering clematis. Scoop out a hollow 10 cm deep. Dip a clematis stem into the hollow and hold in position by pushing in a wooden peg. Return the soil and ram home.

is, the proximal point of stem where it enters the ground – and shorten the stem to 15 or 20cm beyond its emergence from the ground and then lift what remains, which is your new plant.

You can, as a refinement, plunge a largish pot filled with good soil into the ground and layer into that. Then, at lifting time, there will be no disturbance or damage to the young roots, but these are – at any rate in all the hybrids – almost as tough as boot laces, so not to worry, really. Again, with most layering of shrubs it is important to make a sharp elbow at the layer's lowest point because this checks the sap flow which in turn encourages rooting; furthermore, where the layer makes its exit from the ground you encourage, by staking it, an upright habit so that the young plant is a good shape and not lopsided. Well, you can't have a lopsided clematis. The plant being flexible can be led in any direction and the elbow isn't necessary to rooting either. You do want to be sure that your layers keep nice and moist throughout the summer, but then that applies equally to the parent plant, whether it has been layered or not. Adequate moisture in the growing season is an axiom in clematis cultivation.

Division

Herbaceous clematis like *C. recta, C. heracleifolia* var. *davidiana* and *C. integrifolia* can be propagated in the dormant season by simple division.

If you have to lift an old specimen of a large-flowered hybrid you will often find that it has a number of stems arising from below ground level. I have known as many as ten or twelve with 'Nelly Moser', and this condition is encouraged by deep planting in the first place. All these shoots will have roots on them below the soil, and although attached one to another, they are easily cut apart so that a series of ready-made new plants will be the result.

Chapter 10
HYBRIDIZATION

Unless deliberate improvements by hybridization are being sought, the raising of clematis from seed is seldom worthwhile. With large-flowered hybrids, never were so many geese born in the guise of swans, as regarded by their proud raisers. Unless you set about a deliberate policy of hybridization and are extremely critical of the results, you would do best to leave them well alone.

That was the dire warning from my last edition. It is obvious that a number of enthusiasts *are* busy, and it is true that a number of successful clematis have been produced by amateur raisers. The list includes 'Dr Ruppel', *C. alpina* 'Burford White', 'Niobe', *C. alpina* 'Frances Rivis' and *C. montana* 'Freda'. Few commercial growers spare the time and labour to experiment, and they tend to leave the field to the amateur. In such a book as this, we felt that some advice on the techniques used should be included, although you will still, I fear, find that evaluation of your creation will be the most taxing test of all.

Hybridization is possible between virtually all species in the genus. This may sound exciting, and in the case of *C. texensis* crossed with large-flowered hybrids, for example, the results justify the effort. In others I am not so sure. At an RHS show in 1974 the results of crosses between *C. macropetala* and *C. montana* were worthless, although they undoubtedly betrayed their parentage, evidencing characteristics from both species. The first practical snag is that some cultivars set seed more easily than others; *C.* × *jackmanii*, and some of its derivatives such as 'Ville de Lyon', for example, are habitually sterile. *C. texensis* will set seed but is very reluctant. Of the viticellas, 'Minuet' will produce bucketsful without prompting but is the only one in the group that will, with the exception of some strains of *C. viticella* itself.

You need to take notice of some more ground rules. Crosses of two cultivars will show a huge variation, possibly showing only a remote resemblance to the parents, favouring distant (and forgotten) ancestors more strongly. This means that many variations may exist, and you must sow as many seeds of a cross as possible if you are to find the best available result. This is time- and space-consuming; only the most dedicated and patient get far.

Hybridization may not only alter flower colour, it can also affect size. Most often, the results overwhelmingly go opposite to what most people want. Most offspring are runts. Horticulturally speaking of course, so no letters please! Anyone who has sown a handful of *C. orientalis* 'Bill Mackenzie' seed will know what I mean. The pathetically poor flowers that result make the quality of the parent irrelevant.

All is not doom and gloom. Given a large enough sowing, some offspring may be marginally larger than their parents. If these are selected and the best crossed with each other, dramatic improvements in size can sometimes be achieved. This is precisely what Max Leichtlin did with *C. texensis* (described on pp. 74–5). The result was a flower three times larger than the wild plant with which he had started. To maintain quality, we must, of course, thereafter propagate solely from cuttings. Sow seed and we're back to runts!

Three methods of hybridization are open to you. In the first, you simply plant any old seed that happens to set on your favourite clematis. This could have crossed with anything, but no matter. For the person who wants to 'try it and see' without fuss, it is as good a method as any. A number of clematis on today's lists resulted in this way, 'Allanah' and 'Snow Queen' to name but two. As a minor variation to this theme, the parent clematis may do even more of the work for you by shedding seed, unnoticed, which germinates as 'chance seedlings' in the garden. The species and small-flowered hybrids are more obliging in this; *C. montana* seed, for instance, sometimes germinates in the garden like mustard and cress.

The second method gives you a little more control but needs more effort on your part. In this, you choose which plants you wish to cross. When flowers on both are fully open you snip a bloom off one and press the centres or 'eyes' of them both together. 'Rubbing noses', someone succinctly termed it. Label the flower which you expect to set seed with the details of the cross and wait for it to ripen. Bear in mind that this technique is flawed; random pollination from another plant by insects could occur before or after your deliberate cross and you cannot therefore be certain of its pedigree. To overcome this, you must be still more meticulous by adopting a third procedure.

Select your two flowers, as explained above, but, about a week before you think that they might open, decide which is to be the female parent (or seed-bearer). You must then snip away with a pair of fine scissors first the sepals and then the stamens. The latter carry the pollen or male gametes which could, if left in place, cause self-pollination and ruin your cross. At this stage in the development of the flower, the anthers at the end of the filaments will not have begun to release their pollen. They should look waxy, smooth and most probably green in colour. If they are shedding pollen grains, then abandon your attempt to cross. Self-fertilization may have already taken place.

Having decided that all is well, you must then cover your mutilated flower to prevent the ubiquitous bees gaining access and bringing unwanted pollen with them. A muslin bag (which allows ventilation) was traditionally used, but fine-meshed nylon is probably now more readily available. Do the same with the other flower, whose pollen you intend to use, although this bloom can remain intact. Advice on hybridization often omits the need to prevent contamination of the pollen in this way, but I should have thought it an obvious point; if a dozen or two insects have been at your pollen source, unhindered, before you make the cross, I fail to see how you can be certain of the result. Check the stigma each day until you notice the end shiny and viscid. This will indicate that it is ready to accept pollen. Transfer this by, again, rubbing their noses together, or via a fine-haired brush. Either way, once this has been done, replace the muslin bag, and label the cross.

Some hybridizers go back a day or two later and repeat the process as an insurance. Others dispense with the second flower and use pollen collected some weeks or months earlier, having stored it in airtight containers in a refrigerator.

Whatever variations you choose, a peep into the bag a couple of weeks later should show the feathery styles beginning to extend and (if you have been successful) the ovaries at their base starting to swell. At this stage, remove the muslin; your flower is no longer receptive to stray pollen.

Ripening of seed can take months. It will be truncated fatally by the onset of winter if you attempt the cross too late in the season. June flowers of the large-flowered hybrids can usually be relied upon to set viable seed, but not those produced in September. A way round this is to make the crosses on clematis grown in pots in a greenhouse. This will bring forward the flowering period and extend towards winter the time available for development of the seed. It is also the only way to set seed from late flowerers in many parts of this country. It is remarkable just how far you can advance flowering times. Even in a cold greenhouse, you can coax the viticellas, for example, into flower in early May, whereas outside you would have to wait until mid-July.

Sow ripe seed at once. I find it less of a fiddle to leave the feathery seed tails, but remove them if you wish. Take care not to cut into the apex of the seed, for it is from here that the root will emerge. It is advantageous to top off the compost with a layer of fine grit or horticultural sand. This inhibits the growth of mosses and liverwort. The seed pans can be put out in a shady position in the garden, but you will have more control over the unwanted attentions of cats, mice and other pests if you keep them in a cold frame or under the staging of your greenhouse. Keep them moist at all times. Admission of frost should be encouraged, since it often assists in the germination process.

Some species and their close hybrids will germinate rapidly – *C. tangutica* and *C. orientalis* often in less than three weeks. In others eighteen months is normal. Germination of large-flowered hybrids can be frustratingly slow and erratic; a three-year wait is not uncommon. It invariably surprises you when it happens; short spikes of growth appear, reach about 1·5cm and then unfold to give a rosette of leaves. If they germinate in winter (as they often have with me), don't worry. Provided they are not allowed to waterlog, they will take frost without harm. Sometimes, following germination late in the year, the initial spiky shoot will stand still or even wither. Almost invariably, another arises from below at a more sensible time of year; clematis seedlings are very tough.

Some vigorous, small-flowered species will develop fast. *C. tangutica*, if sown in February and treated properly afterwards, will be flowering atop a 1·5m cane, the roots filling a 2-litre pot, by August.

Don't expect your first flowers from the large-flowered hybrids and less vigorous species until at least the second season. If they appear in the first, nip them off. They will not be representative and could fatally weaken the seedling. Remember that if you are growing on under glass, the colour of the flower will not be typical of the same plant grown outdoors.

145

Sports

These are not strictly speaking new hybrids, as they arise from one plant without the introduction of genetic material from another. They are, however, different from the parent plant in that their genetic structure has undergone a spontaneous mutation. Whether this is a permanent change, sufficient to propagate the new cultivar without reversion, is a vital point if they are to be of any lasting value. A parallel situation is that of variegation in some plants, which may show a tendency to change back to the original, non-variegated form.

C. × *jackmanii* cv 'Superba' is the most notable sport. Spotted in a batch of ordinary *C.* × *jackmanii* at the Woking nursery, it has a fuller flower with a reddish tinge when compared to the plainer bluish-purple of the original. Double and albino mutations of *C. viticella* were identified and propagated over three centuries ago. I have never noticed sports on my clematis, but I sometimes wonder how many have been missed among the millions grown, world-wide, over the years.

Chapter 11
CLEMATIS TROUBLES

I think of this as the wilt chapter. Clematis do have other troubles, which we shall examine in due course, but wilt disease is the most serious by so far and away as to eclipse the rest.

Scenario for Wilt

Let me set the scene. An amateur gardener, new to the game, decides to have a flutter with clematis. Although rather shocked at their price he buys three plants, one of them with fat flower buds on it; the others, he is assured, will flower later in the season. With conscientious devotion he follows the book's instructions for planting to the letter, and for the next three weeks he watches his plants' development with mounting excitement. The one with buds isn't making any fresh growth but the buds are enormous and seem likely to burst open at any moment. The other two are making 5cm of growth every day: he keeps them well watered. Then, one evening, immediately on his return from work when he makes his customary inspection, he finds that all the buds on his joy and pride are hanging limply at the neck. Imagine his horror. He syringes it with water but to no avail. Within less than a week the entire plant has shrivelled and turned black. Then one of the other two goes funny. The tip of its fast-growing shoot hangs disconsolately in the midday sun. The rest of the plant looks all right but not for long. That one shrivels and blackens along with the first. Two gone out of three. He feels ill with chagrin and bitter towards the nurseryman.

One of several things can happen after that, but you can see why clematis have earned a reputation for being difficult. They are popular plants, but even so I should think twice as many would be grown if it weren't for the wilt deterrent. It is sad that with almost every florist's flower in which, by careful hybridization and selection, increase in flower power has been attained in triumphant celebration of human skill, every improvement brings in its train dire pests and diseases of undreamed-of potency that had never attended the unimproved plant. Roses have to be sprayed every few days to keep them free of blackspot, mildew and rust; lilies develop the most fearful virus diseases and so do dahlias, the plant shrinking and losing the whole benefit of its acquired characteristics; irises are rotted by smelly bacterial diseases and disfigured with leaf spotting; rhododendrons get bud blast and evergreen azaleas something else that goes for them just as their long-awaited display is getting into its stride in the spring; the potato famine of 1845 was caused by blight disease and so on. Whereas if we stuck to the unimproved buttercups and daisies as we find them in our lawns, we should never have a moment's worry.

Clematis were all right till the breeders started improving them in the middle of the last century. Wilt was already so serious by the early 1900s that Morel, the French nurseryman mentioned so often in these pages, gave up breeding clematis in despair. In 1903 he wrote in the *Revue Horticole* of 'La Maladie Noire des Clématites'. 'A discreet silence has settled on the horticultural press on the subject of this fearful and mysterious disease, but without in any way interrupting its course.' He noticed that it was worst after rain storms and in humid heat, and reckoned that it entered the plant by way of small lesions, wounds, twisted stems etc., and he observed that blackened dead areas developed most frequently at or around soil level. All of which is true, but the disease is not bacterial, as he hazarded it might be.

There is no reference to pests, diseases or any sort of hindrance to the easy cultivation of clematis in Moore and Jackman (1872, revised 1877) but A. G. Jackman (George Jackman's son), brought the subject of wilt, which he called dying off in those days, into the open. He gave it as one reason for the falling off of clematis breeding after 1880. He, too, thought the trouble was bacterial.

The public, however, ascribed it to the nurseryman's wicked practice of grafting (about which they have ever entertained the deepest suspicions), alleging that the vitalba stock overwhelmed the scion and thus caused its death. A. G. Jackman, in a lecture to the RHS in 1916, made the refutation of this theory his main subject, and for the very good reasons that I have discussed in the chapter on propagation. But the public continued to link wilt disease and grafting until very recently indeed. Now that nearly all clematis propagation is done by cuttings and yet wilt continues unabated, that old war-horse has been laid to rest.

Meantime the cause of a stem rot and leaf spot in clematis in the USA was examined by W. O. Gloyer in 1915 (*J.Agric.Res.*, 4, 331–42). The leaf spotting occurred there under glass and we do not find it in this country, but from material showing both symptoms he isolated the fungus *Ascochyta clematidina*. By inoculating it on to healthy plants he was able to reproduce the disease.

That is all that was done on that until our own Glasshouse Crops Research Institute (GCRI) kindly examined the problem, rather at my instigation, in 1965, just fifty years after Gloyer. It was satisfying to find that they, like him, isolated the same fungus from diseased material, *Ascochyta clematidina*, to wit. The work was done by Miss Marion Ebben and Dr F. T. Last and was written up in the GCRI Annual Report for 1965. Their pathogenic isolates from the material examined were more or less entirely restricted to *A. clematidina*. This fungus has also been recorded in Holland, but their nurserymen seem to be even more afflicted by the wilting caused by another fungus, *Coniothyrium clematidis-rectae*.

How the Fungus Works

A. clematidina produces slimy spores and is therefore unlikely to be airborne, but the infected leaf blows about and the fungus can be present in the soil around infected plants, so if you get rid of the plant, the fungus is still likely to be there. Where movable stock is plunged in nursery beds, the vacated area can be treated with formaldehyde diluted at 1 part in 50 and used at 10 litres per square metre. The area must be left vacant for three weeks after treatment or, if you are in a hurry, left for a week and then watered hard. Where pots are stood on plastic sheeting under glass, little trouble occurs.

The fungus may attack any part of the growing clematis down to or just below soil level. One of the commonest points of entry is at or just below ground level, and this causes the maximum damage. The entire shoot above the point of attack collapses and dies, first symptoms showing themselves by the drooping of the young leaves and shoot tip. Within a week, stems and leaves are black. On the other hand, a neighbouring shoot that has not been attacked will continue to grow normally, and the affected stem itself will frequently break again from below the point of infection.

When a stem wilts like this you can always find stem discolourations and/or rotting lesions below the lowest pair of wilting leaves. Experiments showed that the fungus gained entry most frequently where the young stem had been damaged, perhaps by wind or by slugs or perhaps by tying. High humidity also seems to help the fungus to penetrate: wilt often occurs most seriously during growing weather in early summer, as Morel noted.

The fungus also attacks leaves on otherwise healthy plants, and this primary damage is of no importance and will generally pass unnoticed, but infected leaves can act as a source of inoculum spreading the disease to the plant's stems.

When leaves with necrotic (dead) patches on them were collected from nine different clematis species and cultivars growing in different parts of my garden (these plants being apparently quite healthy apart from the odd leaf), *Ascochyta* was isolated from seven of the samples. The appearance of infected leaves varies according to the thickness and texture of the leaf itself. On thin-leaved varieties such as 'Bees Jubilee', 'Marie Boisselot', 'Ville de Lyon' and 'Lasurstern' the dead areas occur in large amorphous patches, the whole leaf often becoming greyish-brown, dry and wrinkled. *C. armandii*, with its leathery evergreen leaves gets discrete (i.e. separate and distinctly marked off) brown areas with a green halo, either as spots or larger more or less oval areas, often with a raised blacker or darker green edge giving the spot an appliquéd look. On 'Victoria' and 'Comtesse de Bouchaud' buff brown patches are concentrically zoned – this is a very characteristic and easily recognized symptom – the centre of the zonation often being at a leaf edge. So too with the thick-leaved 'Mrs Hope', which also has spotting like *C. armandii*.

The GCRI's investigation concluded that *Ascochyta* is one of the major causes of clematis wilt, affecting grafted and ungrafted plants. They suggested that differences in varietal susceptibility might be attributable to anatomical differences, e.g. in thickness of epidermal tissue, or age at which stems form a protective bark. It had, of course, always been recognized by clematis growers that once a plant was mature enough to have formed a tough stem the incidence of wilt was much reduced, and one can see that the fungus would have much greater difficulty in entering the thick stem of a clematis. It might infect foliage or young stems on the distal portions of the plant, but the damage would be slight and localized here, and hardly attract the grower's attention.

It is worth noting, by the way, the difference between this wilt disease, where the fungus does not spread within the host plant but causes a localized infection, and other wilt diseases caused by *Fusarium* and *Verticillium* spp., where the fungus spreads systemically through the host plant. Thank heavens we're spared that situation.

Control

To get rid of the disease on a nursery, where large numbers of clematis are grown together and on their own, should not be too difficult by practising routine hygiene (as in the formaldehyde treatment described) and the regular application of fungicides. At least it is possible to keep the disease at a very low incidence – especially in the controlled environment of a glasshouse, but even in outdoor plunge beds.

In the garden, where clematis are grown among mixed plant collections, it's more difficult. Quite apart from everything else, there is the possibility that *Ascochyta* species capable of attacking clematis may be found on entirely different host plants.

But the main difficulty, as I see it, is that you cannot *kill* a fungus without killing the plant. You can spray the plant or treat it with a systemic fungicide as I shall describe, so as to *prevent* the fungus spores from gaining entry. If the fungus is already on the plant the same treatment can prevent its spread but it will still be there, and if you let up on your protective treatments it can get on the rampage once more.

I'm afraid it is a sad and inevitable fact of life that the longer you grow clematis and the more you love them, i.e. the more of them you grow in your garden, the greater the build-up you will get of wilt fungus spores in your soil and on your plants. Don't think wilt may never come your way. It will.

However, the situation is not nearly as grim or hopeless as it used to be, and we can learn to live with wilt by thwarting its activities with the help of benomyl, the fairly recently synthesized systemic fungicide sometimes sold under the trade name Benlate. Before dilating on its action and use I will draw attention to one possible drawback in benomyl that occurs as a somewhat sinister side-effect: it kills earthworms. To the nurseryman who plunges his pots in or stands them on beds outside, this will be glad tidings, as worms in pots are a dreadful nuisance, blocking up the drainage and often killing the plant by waterlogging. Also it is difficult or impossible to turn a plant cleanly out of a pot in which worms have been working. But worms in the garden are beneficial, and anyway one is always suspicious that if a chemical can kill earthworms its effects on our fauna may not stop there. Still, I shall go on using Benlate until I hear much worse of it.

The advantage of benomyl as a fungicide is that it is translocated to different parts of the plant from within. Therefore, in its application, you don't have to cover all parts of the plant. Use it as a drench to the plant's roots. The chemical's potential for movement within the plant is limited, and you get it most evenly and thoroughly distributed by allowing the plant to take it into its system through its roots.

Benomyl is marketed as a powder in sachets under the trade name Benlate. One sachet should be mixed into a paste, made up to one gallon and applied to the ground around the clematis stems. It does no harm to splash some on to the lower foot or so of the plant, for it is here that wilt is most liable to strike. Activex is often supplied with Benlate but is only added to the mix when it is to be used on the leaves as a spray. A gallon should be enough for one average-sized plant growing in the garden, but half as much would be enough for a youngster in the first two years. Twice as much would be needed for a really large specimen with an extensive root system. The treatment is repeated at monthly intervals from spring (March) to autumn (early October). It has made all the difference to my life and happiness.

The experience of a correspondent who lives near Selkirk is typical of my own and will serve as an example.

'19 March 1973. Dear Mr Lloyd, Some four years ago I was given two "Jackmanii Superba" clematis plants. Every year they have suffered from wilt which began just when the flower buds were forming. Last year, from the time of growth beginning, I sprayed them with Captan every 8/9 days but in spite of this they again suffered from wilt. Is there anything I can do for them?'

I gave him the Benlate recommendation and dosage and wrote to him again at the end of the year, asking for news.

'29 Dec. 1973. I just cannot speak too highly of Benlate as a wilt prophylactic for clematis, and I should have written and thanked you for recommending it. For four consecutive seasons wilt set in just as the "Jackmanii Superba" were about to come into flower and this was in spite of locally recommended treatments.

'Last season, and as soon as the new shoots started coming away, I watered with a Benlate solution both the vines and the surrounding ground up to 15in to 16in radius and repeated every twelve to fourteen days. There was never a trace of wilt to be seen, and the plants continued flowering right on until the end of October. I would be pleased to have a copy of the new edition of *Clematis* when available.' I suppose I ought to give him one.

One tiny point of interest. Benlate is reputed to be perfectly harmless to plants, and this has indeed proved to be the case with just one exception: my large-flowered strain of *Clematis tangutica*, obtained from Jack Drake's nursery years ago. When drenched with Benlate all its foliage shrivels, though the plant is not killed. The same thing happens if it is sprayed with insecticide. It is oddly sensitive. Most probably allied to this effect is that shown by stock plants of all cultivars in a greenhouse. When Benlate is applied to these, the new foliage sometimes shows a reaction. It ends up smaller and curled in on itself. Although subsequent and previous foliage is normal, the afflicted leaves never recover. The plant grows out of it, but the effect can give a commercial grower an anxious moment when he sees it for the first time.

Apart from the protective use of Benlate against wilt, the other, obvious, step to be taken to counter the disease is the removal and burning of diseased tissue. If a shoot collapses, cut it out *below* the point of attack: this will often be below the soil surface.

Which leads me on to another defensive measure. You should always plant so that the bottom 3–5cm of clematis stem is buried. If it subsequently wilts, there will then be a healthy bit of stem left below the lesion and nine times out of ten this will, in due course, throw a new shoot or shoots. Keep the plant watered and wait patiently. I have a 'Nelly Moser' that wilted, getting on for ten years ago, and showed no signs of life for the next year. It then threw a young shoot and has never looked back since. This story was capped by a visitor's: he had an 'Ernest Markham' wilt case and the plant came to life again after three years. But wait; I've not finished yet. More recently still (1975) John Treasure told me of a 'Marcel Moser' plant on his house that was flowering handsomely, having just returned to life after an absence of *ten* years. Patience is the watchword. It is rather remarkable that clematis roots can keep alive and healthy for such a long time without any help from green leaves above ground.

Gardeners in search of a quiet life without crises and palpitations always want to know which clematis they can grow that are immune to wilt. None can positively be said to be immune, but in practice there are some that rarely if ever contract the disease. *C. alpina, C. macropetala* and their cultivars, for instance. The montana group are pretty safe, too, though young plants of *C. chrysocoma* and some of the more special cultivars of *C. montana* like 'Tetrarose' and 'Picton's Variety' can die in their first winter, but never while growing and I doubt if the cause is wilt.

The viticellas can certainly get it, but seldom do. All the large-flowered hybrids, which are the ones to which the new gardener is first attracted, are susceptible.

When a ten-year-old 'Perle d'Azur' or *C. flammula*, for instance, takes and dies, probably in the winter, it will be found that its thick trunk has become quite hollow and has rotted through from the inside. This condition is common to clematis making thick trunks (seldom the montanas) but cannot be ascribed to wilt. Indeed, the thick trunk protects it against the wilt fungus.

Mildew

Powdery mildew can be a very tiresome and disfiguring disease of some clematis in some situations. Especially where they are sheltered, as on a wall or in an enclosed garden or greenhouse, the fungus attacks flowers, foliage and young stems. Growth is crippled, flowers are much reduced in size, and their characteristic colour is so overlaid with grey and white as to be utterly travestied. The fungus does not get into its stride until late June, so early flowers are unaffected. Varieties that flower twice, like 'Lady Northcliffe', may have a normal first crop and a crippled second. Those that are entirely later-flowering are the worst sufferers, if that way inclined – I have never seen mildew on 'Lady Betty Balfour' or 'Mme Baron Veillard'. As special martyrs I would single out 'Ville de Lyon', *C.* × *durandii*, *C.* × *jackmanii* and its close relative 'Superba', 'Comtesse de Bouchaud' and 'Star of India'; also 'Etoile Rose', 'Gravetye Beauty' and 'Sir Trevor Lawrence' in the texensis orbit.

Fortunately the Benlate drench treatment works miracles against mildew, so I need only refer the reader back to Control of Wilt. For control of botrytis on cuttings, please turn to the section on propagation.

Viruses

The incidence and identification of viruses in clematis have never been investigated. Our plants are therefore not officially subject to virus diseases. The evidence of one's eyes nevertheless suggests that they do occur, but as the nurseryman will naturally be inclined to choose healthy-looking material for propagation, any virus or viruses present cannot be giving serious and evident trouble. But they may, without our noticing, be reducing the vigour of a plant.

For many years I owned a plant of 'Mrs Cholmondeley' whose young foliage was invariably mottled with a pale and darker green mosaic, while the outline of the leaves was notably distorted. As the growth matured, the symptoms became masked and the plant grew and flowered reasonably well. Another plant, this time 'Comtesse de Bouchaud', used to develop a green and yellow mosaic pattern on its expanded foliage, following the outline of the veins. I currently have a young plant of *C. rehderiana* showing the same pretty patterning and without its vigour being noticeably impaired. One is somewhat disturbed and I have little doubt that these are expressions of virus disease. Certain old cultivars that used to be noted for their vigour and easy cultivation have mysteriously become difficult with the passing of the years, or if not actually difficult at least they can no longer, in most situations, be called vigorous. One suspects virus as the cause of this weakening.

It doesn't do to worry too much about these things. If a plant or variety becomes weak, we automatically discard it and another, stronger, takes its place. The situation takes care of itself.

Pests

The trouble about any chewing pest is that there is not merely the primary damage to weep over but the fact of the damage letting in *Ascochyta clematidina*, the wilt fungus. Still, if it's around, I suppose it'll get in somehow even if the entire plant is strung with burglar alarms.

Slugs and snails can be very troublesome on newly planted clematis, destroying young shoots at ground level as soon as they appear. If allowed to continue, this can kill the plant altogether. On this subject I shall quote from a letter Hugh Thompson once wrote me. Apropos of wilt: 'My impression is that the most important thing is to stop slug-gnawing at ground level. This damaged area nearly always gets infected by the fungus. I suffer badly from friends and relatives with out of control dogs, so in protection put a sunk collar 6–9 inches high round the base of each clematis. The collar is made of claritex, and I find very few slugs overcome this obstacle, and with very intermittent addition of slug pellets I can stop this ground level damage.'

It is no use thinking you will gain permanent control of slugs with pellets alone. The slugs are there, with the one thought in their minds of eating your clematis, for 365 days in the year, sometimes 366. Your mind is on them only now and then. They'll always have the last chew on you.

Snails are great climbers, and if the clematis is against a wall they may enjoy it at a considerable height. But it is the ground level meals that really matter. Snails do at least hibernate in the cold weather.

Rabbits have been very troublesome on my nursery stock. They are fondest of young leaves and prevented newly potted cuttings of *C.* × *jouiniana* 'Praecox' from making any headway for several months in spring and early summer. (Subsequently we had a merciful recrudescence of myxomatosis.) Other young clematis managed to grow above rabbit range and only had their lower leaves removed, the stems being left intact. In the depths of a northern winter Tom has found rabbits troublesome too, but in a different way. Eighty *C. montana* var. *rubens* had 20cm sections of lower stem removed so neatly that he thought of sabotage with secateurs. However, rabbit prints and droppings in the snow gave the cause. They were highly selective; *C. montana* 'Alexander' and *C. montana* itself were left untouched. Doubtless hares and deer would cause similar trouble only worse. Squirrels have the playful habit of eating off the swelling flower bud just before it is ready to burst open.

Mice are keen. Here is Hugh Thompson again. 'The local very handsome mouse, *Sylvaemus flavicollis,* abounds in my neighbourhood [he lived at Churt in Surrey], and as it has about twice the volume of an ordinary fieldmouse, eats twice as many clematis shoots. *Flavicollis* even killed a long established tangutica by shredding the inch-thick woody stem for nesting material.'

Man himself must be considered a pest within this category, and again I have Mr Thompson to thank for this information. 'I note in a French Encyclopaedia that clematis shoots are greatly esteemed for eating particularly by Italians and Russians! It does not go so far as to suggest that some difficult large-flowered hybrids are the most favoured. The moral appears to be not to have an Italian gardener; nor do I much fancy the prospect of some bulky commissar in a disputed territory sitting down to a meal of macropetala.'

All the same, I do find the young shoots of *C. recta* tempting, and they are so very numerous and strong ... but there, I need further encouragement and information before daring, especially as Mrs M. Grieve in *A Modern Herbal* writes in particular of the bruised leaves and flowers of *C. recta* as irritating 'the eyes and throat, giving rise to a flow of tears and coughing; applied to the skin they produce inflammation and vesication, hence the name Flammula Jovis. They are diuretic and diaphoretic, and are used locally and internally in syphilitic, cancerous and other foul ulcers.' Good Lord deliver us. She continues: '*C.*

flammula is said to contain an alkaloid, Clematine, a violent poison.'
Clematine might be a good name for a poisonous child.

Birds have lately taken to pecking out the swelling buds of the early
flowering alpinas, macropetalas and montanas in my neighbourhood.
At Sissinghurst Castle they have to net their plants in order to obtain
any blooms. For the first time ever I had virtually no bloom at all on
any of my montanas in 1975. I cannot see myself netting them. 'Curb'
and the latest follow-up to 'Curb' (? 'Curb Improved'?) were of no avail
at Sissinghurst in 1975, their head gardeners told me, but the early part
of the year was unusually wet, admittedly.

Be careful that you don't blame the birds for all bud damage though.
The culprit might be one of the winter-feeding caterpillars. Nocturnal
and able to resist frost, they may be shelling out buds in late winter
and early spring while you are abed dreaming of spring's flowers. If
you see damage, try an examination by torchlight after dark; during
the day they hide up. Other insects may be at work; on 27 February, in between cold snaps (admittedly in an otherwise mild winter),
Tom observed numerous earwigs active on his still largely dormant
plants.

Among the small fry, earwigs are the most serious clematis pest.
Their populations build up throughout the summer – more in some
years than in others – and from July onwards they are liable to eat all
young clematis shoots so that nothing remains except the stalks and
midribs, while the flowers are gashed into tatters.

All this at night. Gardeners are mainly diurnal in their habits; if they
go into their gardens after dark it is to saunter and to sniff at their
tobacco plants, not to search for pests. They often fail, in consequence,
to identify the real cause of the trouble. Yet it is only necessary, by day,
to pull aside any folded or overlapping sepals to find the culprits snugly
roosting, while torchlight will reveal them dangling in hordes as they
revel in their midnight feasts.

DDT, when we were allowed to use it, was the most effective spray
control for earwigs. Nowadays any BHC preparation is likely to be as
(in)effective as any other. Just follow the maker's instructions and blame
yourself if the results are nil.

Aphids (greenfly) can be a nuisance on the softest extension growth
of young clematis shoots, especially in May–June before the aphids'
parasites have built up in numbers. When this has happened, aphid

populations will suddenly drop like the temperature in a thunderstorm (and a sudden thunderstorm can be highly deleterious to aphids, also). I have never found that aphid damage was so serious in the garden as to demand action on my part, but on nursery stock you do have to spray at irregular intervals during the growing season.

Discoloured Flowers

In May, when the first blooms expand on your large-flowered hybrids, you may be amazed to find that they are not at all the colour you expected, not even white when they should have been white, but green. According to your temperament and upbringing you will find this intriguing, dull or repulsive. Hugh Thompson wrote me one August: 'I have found this a very mixed year for clematis, the sunless cool early summer affected some but not all. "Ville de Lyon" and "The President" have been very deficient in flowers and I am sure the same will prove true of "Lady Betty Balfour". The cvs that are liable to green have never been greener. Double green "Countess of Lovelace" in great quantity was very odd – just the thing for a zany flower arranger. "Barbara Dibley" passed direct from green to your "dust and ashes" stage without any development of the petunia pigment. Quite hideous at all stages.'

The next year, 1973, in mid-June I received a letter from a customer in Surrey. 'Dear Mr Lloyd, The "Barbara Dibley" clematis which I bought from you last year has flowers which are very far removed from the one I saw at your place and the description in your catalogue. Instead of being petunia red they are a sort of greeny white with faint purple bars. Rather a boring colour, in fact . . . Could the strange colour be due to the soil, the position or might the plant have been wrongly labelled? I'd be grateful for your comment as it's disappointing as it is.'

I wrote her that deprived plants were liable to this condition in their *early* blooms, especially when the spring had been cold and if they were short of water. But I assured her that matters would right themselves. She replied a few days later that the subsequent blooms were getting progressively nearer the colour she had expected.

The double white 'Duchess of Edinburgh' is also very liable to this greenness and one must, I think, bear in mind the fact that the division, in clematis, between leaves, bracts and sepals is often not hard and fast. The trouble is sometimes ascribed to potash deficiency, on the same principle, I take it, as greenback in tomatoes being known to have lack of potash as its cause. No experimental work has been done on clematis to determine whether this theory is correct, nor is it ever likely to be. It could do no harm to give them an extra dose of sulphate of potash and it might help.

Spring Frosts

Nearly all large-flowered cultivars are precocious by nature. No sooner have they stopped one year's growth than they're busy with the next. If you are prone to late frost, hold off cultivars such as 'Vyvyan Pennell' and some of the other doubles which produce their best flowers from last year's wood. Clematis which flower best on their young wood will be a success; frosted shoots can be removed completely without affecting the plant's capacity to flower later on.

Damage by spring frosts can be especially regular and depressing to the gardener (especially as he frequently fails to understand what has happened) when the montanas, alpinas and macropetalas have been grown in an exposed position. These are such hardy clematis that they are frequently not coddled by allocating them a cosy wall. On the contrary, they are expected to cover garage roofs, which are open to the skies, and to deck trees and shrubs which are often pretty exposed themselves. If early warmth unfolds the protective leaves in which the flower buds were nestling, a subsequent sharp radiation frost can polish off the lot. But as the buds were small and colourless, the gardener doesn't notice that they've been blackened, at the time. Later on, when no flowers appear, he ascribes their absence to a bad strain of clematis or to his incorrect pruning. But I don't think you can really be expected to fend off all these acts of God (except by touching wood, of course). Next year is sure to be better.

Chapter 12
DESCRIPTIONS OF CLEMATIS
IN CULTIVATION

In Chapters 5 and 6 we discussed and compared a large number of clematis. This chapter should be used in conjunction with what we wrote there, but it also includes brief notes on many more species and cultivars not mentioned elsewhere. Either they are new to us or unfamiliar to us or they strike us as of minor horticultural importance. We are, therefore, recording rather than passing judgement. For this revision, we have tried to provide more information about cultivars being offered by major growers. Of those which we have seen, we offer an opinion. The others will be allowed to bask in the radiance of descriptions given by their breeders and sellers.

On the clematis that we do know well we have written notes in some detail, having stood or sat in front of examples of each of these, armed with pen and paper and a ruler. We hope these efforts may help readers who are trying to sort out questions of identification. There is, after all, a great deal of confusion in naming, and not much can or should be taken for granted.

No doubt we shall be adding, unwittingly, to the confusion ourselves, at times. Some of our descriptions are derived from clematis grown in pots or under glass. Their performance and appearance will not then be characteristic of their behaviour in the garden. This is especially true of leaf and flower size; also of flower colour.

We have not, in most cases, attempted to give times of flowering, as these can vary wildly according to where you live, how you pruned (or didn't prune) your plant, and the nature of the season from one year to the next. In an early year *C. armandii* will flower in many gardens in January and February, whereas March–April might be considered normal, and May in a cold year.

You can get the relative times of flowering from the letters A, B or C appearing in brackets after the name of each clematis: (A) early, (B) mid-season, (c) late. These letters also refer to pruning methods as given in Chapter 8. Where (B or C) is given as an alternative pruning method it also means that adopting the B method you will get an early crop of blossom followed later by more flowers on the young growth, whereas using the C method of hard pruning, the entire concentration of blossom will come late.

'Abundance' (c). Vigorous. Leaflets 5, and further subdivided. Flowers semi-nodding, 5 cm across. Sepals 4 or 5, pink-red with darker veins. Stamens creamy.

C. addisonii (c). Viorna group from S.E. United States. Height 30–45 cm. The leaves are glaucous and mostly simple. Flower characteristically urn-shaped. Sepals 4, their tips slightly recurved. Rosy-purple outside, cream within with an edge of cream extending to the outside margin. Anthers cream.

C. aethusifolia (c). N. China 1910. Deciduous climber. Attractive parsley-like leaf, finely divided. Flowers held erect on 2 cm stems, cream-yellow bells recurved at tips. 4 sepals, 2 cm long. Seen growing by Raymond Evison alongside the Great Wall of China; he tells me that it has a strong cowslip scent.

C. afoliata (syn. *C. aphylla*) (A). New Zealand, North and South. Introduced 1908. Climber to 8ft or more. Evergreen in a special sense; no true foliage, the leaf being reduced to a tendril. Stems rush-like, green, becoming yellowish in winter. Flowers in clusters of 2, 3 or 4, nodding, 2–2·5 cm across at mouth. Sepals normally 4, 3·5 cm long, obliquely spreading, pale-straw-coloured with green undertone. Dioecious. An untidy grower, usually scrambling over rocks in the wild. Pleasing in flower, it should have a sweet daphne-like scent. Mine didn't.

C. akebioides (syn. *C. glauca* var. *akebioides*. *C. orientalis* var. *akebioides*) (c). China. 1924. Similar to *C. tangutica*. Vigorous to 5 m. Leaves glaucous, pinnate, 5–7 leaflets. Pendant flowers yellow or greenish-yellow, tinged on outside with green, bronze or purple. I am not personally familiar with this.

'Alba Luxurians' (c). Vigorous. Leaflets 7, further subdivided. Flowers

9cm across. Sepals 4, white, with green, recurved tips. Pronounced eye of dark stamens.

'Alice Fisk' (B). 'Lasurstern' × 'Mrs Cholmondeley'. Pale 'blue' flowers 15–18cm across. Usually 8 sepals, rough-textured, pointed and with broad ribs. Pale filaments and slightly boring purple anthers. Pretty when young but fades badly, causing the whole plant to look a mess. (Tom dissociates himself from this description!)

'Allanah' (C). Raised in New Zealand by Alister Keay, introduced into UK in 1984 by Jim Fisk. Vigorous though compact. Very dark red like 'Mme Grangé'. 8 sepals, dark brown stamens. Might prove good. Alister Keay wrote Tom that it arose as a chance seedling and he cannot therefore provide a pedigree.

C. alpina (syn. *Atragene alpina*) (A). Europe to N.E. Asia. 1792. Deciduous climber to 3m. Flowers nodding. Sepals 4, 3–4cm long. Whitish petaloid stamens. Var. *sibirica*, introduced from Siberia 1753, has yellowish-white flowers. For named cultivars of these two see under 'Burford White', 'Columbine', 'Frances Rivis', 'Frankie', 'Helsingborg', 'Jacqueline du Pré', 'Pamela Jackman', 'Rosy Pagoda', 'Ruby', 'Willy', 'White Columbine', 'White Moth'.

'Andrew' (B). From Magnus Johnson, a 1952 seedling of 'Prins Hendrik'. Moderately vigorous to 2m. Flowers 15–20cm across, 8 sepals, bluish-violet, purple stamens.

'Anna' (C). Also Magnus Johnson's. 1974, from 'Moonlight'. Rarely offered in UK, strong grower, leaves ternate. Flowers 12–15cm across. Bicolour: purple on rose. We include this, and others of Magnus Johnson's, solely on descriptions given in the International Clematis Society *Journal*, No. 1, Spring 1984.

'Annabel' (B). A cv from Pennell's with 'Beauty of Worcester' as the seed parent (always a good bet) pollinated by 'Marie Boisselot'. A well-presented flower. Six sepals of a lightish blue with cream stamens. Not very different from other, established blues.

C. apiifolia (C). China. Vigorous to 4m. Flowers creamy-white, 1·5cm across. Vitalba-type. Not very exciting.

C. armandii (A). China. 1900. Vigorous evergreen climber to 6m but requiring wall protection. Leaves ternate, large and tough; coppery when young. Flowers 5cm wide in dense axillary clusters. Delightful scent. Sepals, 5 or 6, white. Stamens creamy. Several clones shelter under the pleasing names of 'Snowdrift' and 'Apple Blossom'. 'Jeffries' Form' has narrow leaflets. Smallish flowers, well branched (multifid) and in dense clusters.

C. × *aromatica* (syn. *C. coerulea-odorata*) (c). (*C. flammula* × *C. integrifolia*), circa 1845. Semi-herbaceous, non-clinging, to 2m. Leaves sometimes simple, ovate or irregularly lobed; sometimes compound, with 3 or 5 leaflets. Panicles of 4cm-wide, violet-blue flowers with white stamens. Scented.

'Asao' (B). A recent introduction from Fisk's. Originated in Japan. Large-flowered, 6/7 sepals, somewhat reflexed. Sepals fairly broad, coming to obtuse-angled apex. Rosy-carmine, fading to off-nothing in centre. Lacks personality.

'Ascotiensis' (c). Vigorous. Leaves simple or ternate. Flowers 13cm across. Sepals 4, 5 or 6, ovate, 5cm wide. Bright lavender-blue. Stamens greenish.

C. australis (A). A New Zealand species. Very low climber, slender and much-branched. Lampshade-shaped green flowers 2·5cm across, sweetly scented. A variable species, Joe Cartman writes that it is a form of *C. forsteri* (Canterbury Botanical Society NZ *Journal*, No. 20, 1986). As with the other New Zealanders, the male has the larger flowers.

'Barbara Dibley' (B). Moderately vigorous. Leaves ternate. Flowers 15–18cm across. Sepals 8, long, narrow and fine pointed. Sumptuous petunia red throughout. Fades badly. Anthers reddish-purple. One of Rowland Jackman's crosses.

'Barbara Jackman' (B). Moderately vigorous. Leaves ternate. Flowers up to 15cm across. Sepals 8, broad, overlapping, tapering briskly to sharp points. Bluish-purple with vivid magenta bar, at first; fades to pale mauve but central bar remains crimson. Contrasting eye of creamy stamens.

C. barbellata (A). A Himalayan version of *C. alpina*. Cv 'Betina', as I have seen it, is a mahogany (beetroot) coloured clone. The Swedes have crossed it with *C. alpina sibirica* and *C. koreana*, but I have not seen the results.

'Beauty of Richmond' (B or C). Vigorous. Leaves simple or ternate. Flowers 15cm or more across. Sepals 6, rosy-blue. Anthers creamy.

'Beauty of Worcester' (B or C). Compact grower. Leaves simple or ternate. Double flowers on old wood; single on young. Sepals 6 on single blooms, straight edged, overlapping, tapering to sharp points. Well-formed, solid flower of rich, deep blue with slight reddish flush at centre base of sepals. Distinct creamy stamens, with tinge of green.

'Bees Jubilee' (B). Weakish. Leaves trifoliate. Flowers 15–18cm across. Sepals usually 8. Colouring as for 'Nelly Moser' but more intense and vivid.

'Belle Nantaise' (C). Moderate vigour. Leaves simple or ternate. Flowers 20cm across. Sepals 6 or 7, tapering to acute points. Pale lavender with whitish stamens.

'Belle of Woking' (B). Moderately vigorous. Leaves simple, broad and rounded. Flowers very double, 10cm across, silvery-mauve.

'Blue Bird' (A). *C. macropetala* × *C. alpina*. Raised by Dr F. L. Skinner of Dropmore, Canada. Vigorous and prolific. Sepals long and twisted, rather similar to 'Rosy O'Grady' but slaty blue. A good sprinkling of repeat flowers through the summer.

'Blue Boy' (C). *C. integrifolia* × *C. viticella*. Dr F. L. Skinner of Dropmore, Canada, made this cross in 1947 and it has recently been evaluated in the ICS *Journal* (No. 4, 1987), by Magnus Johnson. Romke van de Kaa is enthusiastic about this variety which, we are told, grows to 2·5m with hyacinth-blue flowers shaped like open bells, 5–8cm across. We hope to have it soon. The same cross, made many years previously, resulted in *C.* × *eriostemon* and *C. integrifolia* 'Hendersonii'.

'Blue Gem' (B or C). Leaves simple (usually) or ternate, cordate or ace-of-spades-shaped, *not* drawn to fine points. Buds slender, woolly. Flowers up to 15cm across. Sepals 7 or 8, overlapping with blunt, wide-angled points. Light lavender-blue, somewhat darker at centre-base of each sepal; veins throughout darker than background. Stamens fairly conspicuous; filaments white, anthers purple. Not much character.

'Blue Giant'. See 'Frances Rivis'.

'Bonstedtii Crépuscule'. See *C. heracleifolia*.

'Bracebridge Star' (B). Weakish. Leaflets 3, widely separated on long sub-petioles. Flowers 13cm across. Sepals 8, narrowish, inclined to be gappy. Mauve with deeper rosy-mauve bar. Anthers rosy-mauve.

C. brachiata (C). South Africa. Vigorous climber, more or less ever-green. Leaves pinnate. Leaflets usually 5 with large, coarse teeth. Flowers in axillary panicles (with up to 30 in each leaf axil) throughout the terminal 0·8m of every current season's shoot. Flowers 4cm across, sepals 4, occasionally 5, ovate lanceolate, spreading or slightly recurved, green-tinted white. Stamens prominent, greenish-yellow. Scent very slight, of a pleasing vegetable quality. Hardiness not properly tested, yet.

C. buchananiana (C). Similar, but inferior to *C. rehderiana*. Strong grower. Hairy stems and leaves. Tom has grown it for five years without getting a flower; John Treasure had precisely the same problem at Burford (Shropshire). He slung it out.

'Burford White' (A). A clear white selection of *C. alpina*. Pale foliage. Grows to 2·5m. John Treasure says that it was a chance seedling, raised by one of his customers.

C. calycina. See *C. cirrhosa* var. *balearica*.

C. campaniflora (C). Portugal. 1820. Vigorous, climbing to 4·5m. Leaves pinnate with 5, 7 or 9 leaflets, subdivided into threes. Small, 3cm-wide bell-flowers. Sepals 4, bluish-white.

'Campanile'. See *C. heracleifolia* 'Campanile'.

'Capitaine Thuilleaux' (B). A 'Nelly Moser' type clematis from France with 8 rounded sepals. Broad vivid carmine bar against narrow pale margins.

'Cardinal Wyszynski' (C). Raised in Poland by Brother Franczak and introduced by Jim Fisk in 1986. Described in his catalogue as 'glowing crimson'. Too early to evaluate.

'**Carmencita**' (c). From Magnus Johnson, a 1952 seedling of *C. viticella* cv 'Grandiflora Sanguinea'. From his description, it is vigorous to 3m and resembles 'Kermesina' but is deeper and without the paler centre. Stamens deep purple.

'**Carnaby**' (B). Introduced by Treasure's from the USA in 1983. 6 sepals, wavy-edged, recurving slightly at tips. A deep raspberry-pink with a deeper bar, reddish stamens. Compact and free-flowering. It has not featured in many growers' lists recently. John Treasure says it fades rather badly and has been overshadowed by 'Dr Ruppel'.

C. × *cartmanii* 'Joe'. Natural cross between *C. marmoraria* and *C. paniculata*. From Joe Cartman, a keen NZ gardener. Cool greeny-white flowers, 6/7 sepals 4cm across. Very showy and prolific. Cut leaves, evergreen, no scent. We saw this in flower, under glass, at Jack Elliott's (Kent) on 28 March 88. Delightful.

'**Cassiopeia**' (B). 1952 seedling of 'Prins Hendrik'. Raised by Magnus Johnson, we include it on the strength of his description. 'Medium vigour up to 2m. Leaves ternate. White flowers up to 17cm across tinged with purple when young. Flowers quite freely June to Sept.'

'**Chalcedony**' (B). As far as we can tell, offered only by Peveril's. They describe it as 'double at all times, ice-blue and a strong plant'. No comment; we have tried unsuccessfully to find out more from them.

'**Charissima**' (B). Seen at Chelsea in 1984 on Jim Fisk's stand. A raspberry colour all through, with reddish stamens and a large flower, 20cm or so across. According to the catalogues it should be cerise-pink with a deeper bar. Maroon stamens. Peveril Nursery say it is sweetly scented.

C. chiisanensis (c). Korea. 1917. Re-introduced by the Scandinavians in 1976. Similar to *C. koreana*.

C. chinensis (c). China. Vigorous to 8m. Leaves pinnate, 5 leaflets. Small white fragrant flowers up to 2cm across. Though previously offered by Treasure's of Tenbury, we have not seen this clematis.

C. chrysocoma (A). Yunnan. 1884. As we know and grow it in this country, a vigorous climber to 9m. Leaves trifoliate, purplish when young, both surfaces covered with down as also are leaf and flower stalks. Leaflets often as wide as long, giving entire leaf a broad look quite distinct from *C. montana* in which the leaflets are lanceolate. Leaflets in *C. chrysocoma* usually 3-lobed, merely toothed in *C. montana*. The 2 side-leaflets are about half the size of the central one. Flowers in axillary clusters of 3–5, 6cm across on extra-long, 8–10cm stalks. Sepals 4, broad and rounded, soft pinkish mauve. (Others call them pink, but this is a question of how you see and define colours.) No scent. The 'true' *C. chrysocoma*, shown at Chelsea in 1984, was very like a white montana. More recently (1987), an expedition to Yunnan brought back more seed which has been widely distributed. We await the variations with interest.

C. chrysocoma cv **'Continuity'** (A). Was given to John Treasure by Rowland Jackman. A good pink form with flowers on long stalks. Continues sporadically through the growing season. Probably not fully hardy. Often listed as plain *C.* 'Continuity'.

C. chrysocoma var. **sericea** (syn. *C. spooneri*) (A). Vigorous climber to 8m. Leaves ternate, leaflets only a little less broad than long, downy on both surfaces, markedly serrate. Flowers bold, 8cm across. Sepals 4, broad, obovate (4cm long by 3·5cm across), white but slight pinkish-mauve flush along sepal midribs. No scent.

C. cirrhosa (A). Southern Europe, Asia Minor, 1596. Evergreen climber to 4m. Leaves simple, ovate or three-lobed with serrated margins. Flowers nodding, 4cm across, each subtended by a pair of green perfoliate bracts. Sepals 4, greenish white. Mainly winter-flowering. Bean describes the seed vessels terminated by plumose styles, forming large, beautifully silky heads. I had never seen these until recently in a Dorsetshire garden. Seen in mid-February, the seed heads were just reaching the silky stage, having flowered well before Christmas.

C. cirrhosa var. **balearica** (syn. *C. balearica*, *C. calycina*) (A). Mediterranean islands etc. Introduced 1783. Besides what I have already written on this clematis (see pp. 58–60) it is worth repeating that with *C. cirrhosa* itself there are many confusing variations. It is known as the fern-leaved clematis, but the *C. cirrhosa* var. *balearica* that received an AM in 1974, and that is in most general cultivation, is not nearly as cut-leaved or ferny as other clones I have seen. A form of *C. cirrhosa* sent me by Mr Fisk has greeny-cream bells without freckles and is apt to flower in summer as well as in winter.

C. cirrhosa cv **'Wisley'** (syn. 'Wisley Cream') (A). Newish. Good-sized cream flowers. No freckles. Leaves paler than *C. cirrhosa*, with minimal serration; shiny, bronze well in winter.

Treasure's also have some cirrhosas which are so heavily freckled as to appear predominantly red. They were produced from seed gathered in the wild in Majorca. It is too early to tell whether they will be suitable for growing in the UK.

C. colensoi. See *C. hookeriana.*

C. columbiana. See *C. verticillaris.*

'Columbine' (A). A selection of *C. alpina*. Flowers 8cm across when wide open, but sepals usually obliquely spreading. Light blue. Staminodes white.

'Comtesse de Bouchaud' (C). Vigorous but not tall, 3m. Leaflets usually 5. Flowers 13–15cm diameter at start of season, smaller later. Sepals 6, markedly channelled along midribs, margins and tips recurved on opening but becoming incurved with age. Bright pinky-mauve. Stamens cream. An abundant flowerer.

C. connata (C). Rehderiana type. Perfoliate, broad base to leaf. I can't get excited enough about this one to move it from my frameyard into the garden.

'Corona' (B or C). Bred by Tag Lundell of Sweden, a seedling of *C.* 'Nelly Moser', and introduced by Treasure's in 1972. They describe it as 'purple suffused pink with an orange highlight', which sounds theatrical. My own note from seeing it at Chelsea is: sepals 8 typically. Coloured petunia-red like 'Barbara Dibley' on opening but sepals blunt-ended. Filaments pale, anthers dark reddish. Not unlike 'Crimson King'. Vigour said to be moderate.

'Côte d'Azur'. See *C. heracleifolia* 'Côte d'Azur'.

'Countess of Lovelace' (B). Tricky to start but can be vigorous enough when established. Leaves ternate. Flowers 15cm across, packed with approximately 70 narrow, pointed sepals. Lilac-blue, fading. Stamens whitish. Carpels form a greenish eye. Light crop of single blooms on new wood.

'Countess of Onslow' (c). Probably no longer in cultivation.

'Crimson King' (B). Rather weak grower. Leaves simple or ternate. Flowers up to 13cm across. Sepals 5, 6 or 7, rather gappy and recurving along margins with age. Fairly uniform but tired crimson lake. Filaments white, anthers markedly brown.

C. crispa (c). South-eastern United States. 1726. Semi-woody climber to 1·5m. Leaflets 3, 5 or more. Flowers solitary, bell-shaped, 4–5cm long, scented. Sepals 4, pale blue, not unlike *C. heracleifolia* var. *davidiana* with recurved points but more solid. Related to *C. texensis* with which it has been crossed. Treasure's used to list a white form which sounded interesting.

C. × *cylindrica* (c). A cross between *C. integrifolia* and *C. crispa*, the hybrid sets seed which has been offered by Chiltern Seeds, Bortree Stile, Ulverston, Cumbria. Described as herbaceous, 1m tall, bluish flowers with thick recurving sepals. *C.* × *cylindrica* 'Nana' was raised by Magnus Johnson.

'C. W. Dowman' (B). Moderate vigour. Leaves ternate with rather long narrow segments, distinctly boat-shaped. Sepals 8, ovate, sharp-pointed, narrow at base with gaps. A lively pink at first but soon bleaching. Anthers beige.

'Daniel Deronda' (B or c). Moderate vigour. Leaves simple or ternate, bronzed. Flowers semi-double, flattish, 18cm across, bluish purple. May miss out on these and go straight to single flowers later, with paler centre to sepals. Cream stamens.

C. davidiana. See *C. heracleifolia* var. *davidiana*.

'Dawn' (B). Habit compact though free-flowering. Leaves ternate (occasionally simple) on long petioles. Leaflets broadly cordate with purple margins. Sepals 8, rounded, bluish white, pinkest near tips, fading to white throughout but contrasting purple anthers. Slight scent. Introduced by Treasure's (originally as 'Aurora') 1969.

C. douglasii var. *scottii* (c). Herbaceous, non-clinging, 1m tall. Leaves glaucous, bipinnate, the leaflets narrow, lanceolate. Petioles and underside of leaves downy. Flowers urn-shaped, 3·5cm across, 4cm long; sepals 4, recurved at tips. Outer surface downy, lavender-coloured, a pinker tinge within. Masses of creamy stamens, slightly protruding. Cv 'Rosea' is a good shade of pink.

'Dr Ruppel' (B). Raised by the eponymous doctor in Argentina and introduced by Jim Fisk in 1975. A rather nasty 'Nelly Moser' type, though Tom and others like it. Only slightly paler at the margins when I saw it, otherwise carmine. Distinctive wavy-edged leaf. Autumn flowers differ from those of the spring in having less differentiation to the sepal margins. Moderately vigorous.

C. drummondii (C). An American species found in dry conditions from Texas to Arizona. Leaves pinnate, coarse, sometimes toothed, finely pubescent or almost glabrous. Flowers white, 2cm across. Though not seen in the UK, it is offered in some Continental lists.

'Duchess of Albany' (C). Vigorous semi-woody climber to 3m. Leaflets 5. 4cm-long bell-flowers. Sepals 4, nearly pink. Sometimes shy flowering.

'Duchess of Edinburgh' (B). Moderate vigour. Leaves ternate, a yellowish green. Flowers double, white rosettes, 13cm across, often misshapen and green. Stamens creamy.

'Duchess of Sutherland' (B or C). Sometimes difficult to start, but vigorous when happy and established. Leaves ternate. Light pruning allows small crop of flattish double flowers, 15cm across. Main crop on young wood later. Sepals 6, very broad and widely overlapping, tapering to sharp right-angular points. Soft rosy-red. Filaments creamy, anthers reddish-purple.

C. × durandii (*C. × jackmanii × C. integrifolia*) (C). Sub-shrub to 2·5m. Leaves simple, ovate, non-clinging. Flowers 10cm across, rich indigo-blue. Sepals 4, 5 or 6, with deeply furrowed midribs. Stamens off-white.

'Edith' (B). Introduced by Treasure's in 1974, a seedling of 'Mrs Cholmondeley'. White flowers with red anthers. Similar in other respects to its parent.

'Edouard Desfossé' (B). Only moderate vigour. Leaves simple or ternate. Flowers 15cm across. Sepals 8, long, rather gappy; deep mauve-purple, fading. Reddish-purple central bar. Conspicuous reddish-purple anthers.

'Edward Prichard' (C). Herbaceous, 1 to 1·5m, non-clinging. Leaflets 5, deeply toothed. Flowers in terminal and axillary panicles, cruciform, 4cm across. Sepals narrow, wedge-shaped, pale mauve, deepest near their blunt tips. Picks well, its pleasant scent filling a room. From a cross between *C. recta* and *C. heracleifolia* var. *davidiana*.

'Elizabeth Foster' (B). 'Vyvyan Pennell' × 'Nelly Moser'. Although sown by Pennell's in 1960, it was only named and introduced by them in 1975. A vigorous bicolour, pink with carmine bar, it is not widely available.

'Elsa Späth' (B or C). Moderate vigour. Leaflets 3 or 5. Very large, floppy flowers, 20cm across. Smaller blooms of better form, later. Sepals 6, 7 or 8, broad and overlapping, rich lavender-blue. Anthers reddish-purple. Sometimes sold as 'Xerxes'; in Australia as 'Blue Boy'.

'Empress of India' (B). An old *C. lanuginosa* type, raised by Jackman. Spingarn describes it as 'light violet purple with deeper purple bar'. Lost for many years, recently re-introduced from USA. Large flowers on old wood but much deeper reddish-purple, on young. 6 sepals, pale stamens.

C.* × *eriostemon (syn. *C. bergeronii, C. chandleri, C.* × *intermedia, C. hendersonii*) (C). Introduced 1835. A cross between *C. viticella* and *C. integrifolia* made at different times in different places. Herbaceous, non-clinging to 2·5cm. Leaflets up to 7. Nodding lantern flowers up to 5cm across, very like *C. integrifolia*. Sepals 4, dusky purple.

'Ernest Markham' (B or C). Vigorous to 2·5m. Leaflets, 3 or (usually) 5. Light pruning (B) allows preliminary small crop of large flowers, 15cm across. Main crop always on young shoots; flowers 13cm across. Sepals 6, broad, overlapping, rough-textured, magenta. Stamens beige, inconspicuous.

'Etoile de Malicorne' (B). A bit of a mystery. Origin unknown and not offered in any commercial lists, as far as we know. Quite a strong grower with bicolour flowers, magenta on lilac. 6–8 sepals, slightly overlapping and nicely shaped, ending in a pronounced point. Reddish-brown stamens. Deserves to be better known.

'Etoile de Paris' (B). Fairly vigorous but compact. Leaves trifoliate, small, dark and leathery. Flowers 15cm across. Sepals 8, fine-pointed, gappy, strong mauve-blue with reddish central bar. Anthers dusty, greyish-purple.

'Etoile Rose' (c). Herbaceous or sub-shrubby climber to 3·5m. Nodding bell-flowers, 5cm long to 5cm wide at mouth. Sepals 4, silvery-pink at margins, cherry-purple-centred.

'Etoile Violette' (c). Vigorous. Leaflets 5. Deep purple, cream-eyed, 10cm-wide flowers. Sepals 6 (sometimes 4 or 5), recurved at tips.

'Fair Rosamond' (b). Vigour only moderate. Leaflets 3 or 5. Fat globular buds opening to full, well-formed flower, 13 to 15cm across. Prominent cushion of bright purple anthers. Sweet scent.

'Fairy Queen' (b). Rather weak-growing. Leaflets 3. Flowers 18cm or more across. Sepals 8, 4cm across, a little gappy. Flesh coloured with rosy-mauve bar. Anthers dusty purple. Previously unavailable, it has recently re-entered the lists.

C. fargesii* var. *souliei (b or c). China 1911. Vigorous to 4m. Young stems purple where exposed to light. Leaves pinnate up to 0·3m long with 5–6cm between pinnae. Pinnae usually ternate, each pinnule tri-lobed and with jagged, uneven teeth. Both surfaces slightly hairy, especially when young. Peduncles 7–9cm. Flowers typically in 3s, arising from one point at which are 2 subtending foliar bracts, 2cm long. Centre flower opens some ten days before subtending buds and one or both of the latter frequently abort. Buds rounded, coming to a short sharp point. Stamens greenish to cream. Sepals 6, white with tinge of green, recurved at margins, which are crimped. No scent.

C. finetiana (syn. *C. pavoliniana*) (b). China. 1908. Evergreen. Leaves simple or ternate, cordate at base, tapering to fine points, 8cm long. Flowers 3–7 in axils, 4cm across, strongly hawthorn-scented. Sepals 4, narrow, white. Stamens and styles white. June. Needs a warm wall. Inferior to the related *C. armandii*.

'Fireworks' (b). From John Treasure. 'Maureen' × 'Nelly Moser'. Similar colouring to 'Mrs N. Thompson'. Sepals long and thin with a pronounced twist giving the impression of a catherine wheel, hence the name. Vigorous.

C. flammula (c). Southern Europe. 1590. Vigorous, to 4m. Foliage rich, dark green. Leaves biternate. Flowers 3cm across in great quantities, powerfully scented. Sepals 4, pure white.

***C. florida* 'Alba Plena'** (syn. *C. florida* 'Plena') (B). Probably the clone introduced by Thunberg from Japan, 1776. Fully double without distinction between sepals and staminodes in colour or form. Greenish-white throughout.

***C. florida* 'Sieboldiana'** (syn. *C. florida bicolor, C. sieboldii*) (B). Weak but capable of reaching 2·5m. Hardy, though early spring growth often cut. Leaflets 5. Flowers borne singly in leaf axils of terminal metre of young growth. Peduncle 20cm long with conspicuous foliar bracts. Flowers up to 10cm across. Sepals 6, ovate, 4cm across; pale green, slowly expanding and maturing to white. Anthers petal-like, massed in a domed boss, eventually 5cm across, changing from green to rich purple and persisting for some time after the sepals have fallen.

C. foetida (A). A species from New Zealand. Found in both North and South Islands, where it is a vigorous grower to 6m or more. Leaves almost glabrous, trifoliate. Free, forming large panicles. Individual flowers are up to 2·5cm across in the male and comprise 5–8 sepals; chartreuse-green and described as strongly scented of boronia. They are most certainly *not foetid*. When I saw the plant at Treasure's in 1984, it reminded me of a muehlenbeckia. Joe Cartman, writing in the *Canterbury Botanical Journal* (No. 20. 1986), says that it has a marked juvenile stage that resembles *C. marata* and lasts 3–4 years.

C. forsteri (A). From New Zealand, where it is a common ingredient of the lowland scrub on North Island. Evergreen, pale green leaves, trifoliate. Can look like *C. paniculata*, but is very variable. We saw a number at Jack Elliott's which had different leaf forms and sizes on the same plant. The male, as usual, has larger flowers which are up to 5cm across. Greeny-yellow with 5 or 6 sepals and yellow stamens. A slight lemon scent. Seed heads fluffy. Quite a vigorous plant, it flowers outside in late May but can be hit by frost in some winters. John Treasure says it is rather nice in the greenhouse but would be quite impossible with him outside (in Shropshire). According to Joe Cartman, the species should now include *C. australis, C. petriei,* and *C. hookeriana.*

'Four Star' (B). Comes from a nursery of the same name in the USA and is a new introduction into the UK by Jim Fisk. His catalogue describes it as pale lavender with a deeper bar. We await the appearance of *C.* 'Diesel'.

'Frances Rivis' (A). The largest-flowered cultivar of *C. alpina* and deep blue in colouring.

'Frankie' (A). A selection of *C. alpina*. Same colour as 'Frances Rivis' but a better-balanced flower.

C. fusca (B). North-east Asia, 1860. Herbaceous or semi-woody climber to 4m. Leaves pinnate, the terminal leaflet often absent. Flowers solitary, urn-shaped. Sepals 4, 2cm long with recurved tips, brown outside, white inside. The peduncle is brown-furred. A collector's piece, interesting as an Asiatic relative of the mainly North American Viorna group. More often than not, *C. fusca*, as seen in nursery lists, is actually *C. japonica*. *C. fusca* var. *violacea* is violet instead of brown, less hairy and inside of sepal is dark.

'General Sikorski' (B). A large-flowered hybrid raised in Poland by Wladyslaw Noll. Introduced by Fisk's in 1980. A compact grower to 2–2·5m. Flowers 10–15cm across, mid-blue with a reddish base to each of the 6 sepals which are slightly overlapping and well shaped. Jim Fisk wrote Tom that it was the product of a chance cross, and no pedigree is available.

C. gentianoides (C). Tasmania. Herbaceous, non-climbing. Dioecious. Simple, ribbed plantain leaves, ovate lanceolate, purple to tip, lighter at base. Bushy habit to 30cm. 1–3 flowers per stem, usually terminal but sometimes lateral. The plant shown at Chelsea in 1986 was female and had 4 sepals, white, slight sweet scent (like *C. recta*). We saw it in Dr Jack Elliott's greenhouse in late March. Lots of flower buds, tinged purple. A good-looking plant.

'Gillian Blades' (B). From Fisk. White with a faint mauve edge. Sepals 8, pointed. Pale stamens. A good flower, rather like a white 'Lasurstern'.

'Gipsy Queen' (C). Vigorous. Leaves ternate. Flowers 13cm across. Sepals 6, broad and rounded, purple fading to violet. Anthers reddish-purple.

'Gladys Picard' (B). From Fisk. A large very pale mauve flower with 8 broad, overlapping sepals and white stamens. Pretty as shown at Chelsea Flower Show.

C. glauca (syn. *C. orientalis* var. *glauca*. Maxim.) (C). W. China to Siberia. Slender climber, leaves glabrous, pinnate, very glaucous. Yellow flowers, campanulate, nodding, 3·5cm across. August to September. Similar to *C. orientalis*, it is allegedly hardier. Recently offered by Treasure's.
var. *akebioides* – see *C. akebioides*.

C. grata var. **grandidentata.** *C. grata* is typically a Himalayan species not in cultivation. This variety was introduced by E. H. Wilson at Veitch's Nursery 1904 and is of Chinese origin. Vigorous climber to 9m. Closely akin and very similar to *C. vitalba*, but said to flower May–June so don't ask the author how it should be pruned. Flowers in terminal panicles, suggesting (C) pruning, but a later flowering season. A correspondent from Leeds, Yorkshire, tells me that his plant flowers on young growth in September following hard annual pruning in winter.

'Gravetye Beauty' (C). Vigorous herbaceous climber. Leaflets 5, often cordate. 7cm-long bell-flowers open out into 9cm-wide stars. Sepals 4, 5 or 6, with incurved margins but recurved tips. Cherry-red, ageing pinker. Filaments cream, anthers red as sepals.

'Guiding Star' (B). Leaves ternate, bronzed when young. Flowers 11cm across, dished. Sepals 6, sometimes 8, very sharp-pointed, bluish purple but not dark. Centre of sepal fades to paler colour than rest, in which there are reddish undertones. Filaments white, anthers brown. Synonymous, we think, with 'Lilacina Floribunda'.

'Hagley Hybrid' (C). So abundant in flower production for three solid months that growth tends to be limited. Seldom taller than 2·5m. Leaves usually ternate. Flowers 13–15cm across. Sepals 6 (or 5), cupped, tapering to fine points. A delightful rosy-mauve at first but fading. Handsome purplish anthers.

'Haku Ookan' ('The White Royal Crown') (B). From Japan. 7- or 8-pointed sepals. Similar violet colouring to 'The President' but with prominent cream stamens. Acceptable.

176

'Helsingborg' (A). Believed to be *C. alpina.* × *C. ochotensis.* Raised by Tag Lundell, flowers deep purple, young stems likewise. A good plant.

'Henryi' (syn. 'Bangholme Belle') (B). Vigour moderate once established, but young plants often fail. Young foliage characteristically bronzed. Leaves simple or ternate. Flowers 15–18cm across. Sepals 6, scarcely overlapping, pure white. Contrasting eye of brown-tipped stamens. Two crops.

C. heracleifolia (C). Eastern China, 1837. Coarse, non-clinging subshrub to 1·5m. Large ternate leaves, the central leaflet twice the size of the other two. Flowers hyacinth-like in size and shape, borne in axillary clusters and whorls. Sepals 4 (5 or 6), blue. Best known in gardens in three named cultivars and one natural variety, but see also 'Mrs Robert Brydon'. *C.h.* var. *davidiana*, Northern China 1864, is entirely herbaceous and clump-forming. Flowers dioecious (usually male in cultivation), pale blue, larger and opening wider (4cm) than in the type-plant. Very fragrant. 'Wyevale' is a slightly deeper shaded cultivar with markedly larger flowers and as strong a scent. *C.h.* cv 'Campanile' is a sub-shrub to 1·5m. Hermaphrodite. Numerous, densely clustered, 2cm-long, mid-blue bell-flowers. 'Côte d'Azur', similar; perhaps, nowadays, the same. 'Bonstedtii Crépuscule' is similar to *C. heracleifolia* but has a smoother leaf and more compact habit.

'Herbert Johnson' (B). A Pennell hybrid: 'Vyvyan Pennell' × 'Percy Picton'. Still listed by Peveril Nursery but, as far as I can see, not elsewhere. 'A very large flower, freely produced. Reddish-mauve with maroon stamens. Quite vigorous.'

'H. F. Young' (B). Mid-blue, nicely shaped flower. Sepals 8, widely overlapping. Stamens white.

'Hidcote Purple' (B). Sepals 7 or 8, rather pointed, reflexed in mature flowers. Light rosy purple. Dark anthers.

C. hookeriana (syn. *C. colensoi*) (A). Leaves pinnate, described as being somewhat fern-like, the pinnae up to 8cm long. Flowers in axillary clusters, star-like, up to 4cm across, scented. Sepals usually 6, green or yellow-green; stamens likewise. A New Zealand species from north and south shores of Cork Strait, ascending to 900m inland. A variable species and sometimes difficult to identify. Introduced by Collingwood

Ingram 1937, and grown in his Kentish garden whence he exhibited it 15 May 1961 and it received an AM as *C. colensoi*. Should be grown on a sheltered wall.

'Horn of Plenty' (B). Leaves ternate with purplish margins. Flowers large, somewhat cupped. Sepals usually 8, symmetrically elliptic, overlapping, rather blunt-tipped. Rich rosy purple fading to a good blue-mauve. Stamens numerous and a prominent feature with reddish-purple anthers.

'Huldine' (C). Vigorous. Leaflets 5. Flowers 10cm across. Sepals usually 6, not overlapping, with slightly incurved margins but recurved tips. Upper surface white; under surface pale mauve, with darker central bar. Stamens greenish-white.

'Hybrida Sieboldiana'. See 'Ramona'.

C. ianthina (C). Korea. A *pitcheri* type, related to *C. fusca*. Flower dusky purplish. Smells of apricots. Spingarn lists *C. janthina* (sic) as a synonym of *C. fusca* var. *violacea*.

'Imperial' (B). Raised by Magnus Johnson, a seedling of 'Lasurstern'. Moderately vigorous. Flowers 10–16cm across, first flush often semi-double. Imperial purple, stamens yellowish. A clematis with which we are not familiar; the description given is from the raiser's article in the International Clematis Society *Newsletter* No. 1, 1984.

C. integrifolia (C). Southern Europe, 1573. Herbaceous to 1m. Leaves simple, ovate with 3 grooved longitudinal veins; non-clinging. Flowers borne singly, nodding, 5cm across. Sepals 4, obliquely spreading, mid-blue on both surfaces, their margins recurved and somewhat twisted. Stamens white, forming a tube. Cv 'Hendersonii' is larger, while cv 'Olgae' has very waved sepal margins and is scented. Cv 'Rosea', at its best, is a strong, quite erect plant. Large leaves. Flowers similar in shape to cv 'Olgae'. Clear pink with a deep lilac reverse. Good scent. A number of inferior clones exist. Peveril Nursery offer cvs 'Pastel Pink' and 'Pastel Blue', which they describe as good companions. The former is, they say, very pale pink, the latter, powder blue.

'Ishobel' (sic) (B). Introduced by Fisk's in 1985. They describe it as 'a large white (15–20cm across) with dark stamens.

Grows to 4m.' Jim Fisk says that it was raised by one of his customers.

C. × ***jackmanii*** (C). Leaves simple (at base of young shoots), ternate (along middle of young shoots) or in fives (towards extremities of young shoots). Flowers 13cm across. Sepals 4 (5 or 6), gappy, rough-textured with channelled midribs. Bluish-purple. Stamens greenish-beige. Cv 'Superba' differs in its broader sepals composing a fuller flower and in the reddish tinge in its purple colouring, especially noticeable in newly opened blooms.

'Jackmanii Alba' (B or C). Very vigorous with pale green foliage. Leaflets 3. First crop of flowers, 13cm across, semi-double, of ragged outline with sepals of unequal lengths and often undecided whether to be leaves, on the outside of the flower, or stamens, in the flower's centre. White with pale blue veins. Second crop, single. Sepals 5 or 6, white. Anthers brown. Not very free-flowering.

'Jackmanii Rubra' (B or C). Fairly vigorous. Leaves ternate. Flowers semi-double from year-old wood, slightly lopsided; single blooms soon develop and take over for the remainder of its long season. Sepals 4, 5 or 6, velvety crimson lake. Cream stamens.

'Jacqueline du Pré' (A). A recent selection of *C. alpina*. Peveril's showed it at Chelsea in 1984, where a Preliminary Commendation (PC) was given. They describe it as 'warm pink outer sepals with soft pink interior, staminodes white, flushed pink'. According to Raymond Evison, it is like *C. alpina* 'Willy'. Peveril's got a PC for *C. macropetala* 'Snowbird' at the same time.

'James Mason' (B). One of Peveril's. They say it is a large white with crimped sepals and maroon stamens.

C. × ***jeuneana*** (A). Thought to be *C. armandii* × *C. finetiana*. Similar to *C. armandii,* but smaller flowers, each 2·5cm across. White, pinkish underside. 6 sepals. Vigorous to 5m.

'Jim Hollis' (B). Vigorous. Fully double blooms without guard sepals. Attractive silvery lavender. One of Pennell's, 1961. 'Barbara Jackman' × 'Daniel Deronda'. Not widely available.

'Joan Picton' (B). Raised by Percy Picton. Free-flowering and a good doer. Sepals 7 or 8, obovate, rosy lilac, purple anthers. Pleasant if not outstanding.

'**Joan Wilcox**' (B). A large pale mauve-pink flower with 8 grooved sepals and crimped margins. Dark purple anthers make a good feature.

'**John Huxtable**' (C). A self-sown seedling of 'Comtesse de Bouchaud'. Late-flowering and translucent white with green stamens. Quite a compact grower. As a late-flowering white, it is a most useful plant. Produced by Jackman's; we believe that John Huxtable, the man, was an employee of that firm.

'**John Paul II**' (B). Another Pole via Fisk's. A cross of unknown parents, it was raised by a monk, Brother Stefan Franczak, and introduced in 1982. 'Flower 10–15cm across, whitish with pink tinges. Reddish stamens. Grows to 5m.'

'**John Warren**' (B or C). Habit, compact. Leaves simple or ternate. Large flowers up to 25cm across. Sepals 6 (or 7) tapering to acute points. Pinkish lilac but richer on opening and poorer on fading, darkest along margins and midrib. Reddish stamens.

C. × *jouiniana* (*C. heracleifolia* var. *davidiana* × *C. vitalba*) (C). Non-clinging sub-shrub to 4m. Leaves dark green, glabrous; leaflets 5 (or 3). Huge, compound, leafy panicles of wide-opening flowers 3–5cm across. Sepals 4, 5 or 6, off-white shading to grey-blue at tips. Stamens cream. September–October, but cv 'Praecox' starts flowering 6 weeks earlier.

'**Kacper**' (B). Another Fisk introduction from Poland, appearing first in his 1987 catalogue. Described as a large (15–20cm) flower, intense violet, sepals crenate, violet stamens. Height up to 4m. Brother Stefan Franczak raised this and named it after his father.

'**Karin**' (B). A seedling of 'Lasurstern', raised by Magnus Johnson. His description is of a moderate grower to 2m. Leaves ternate, ovate, dark green above. Flowers up to 16cm across. 8 sepals overlapping, violet stamens dark purplish red. First flowers sometimes semi-double.

'**Kathleen Dunford**' (B). Introduced by Fisk's in 1982. Named after the New Forest lady who raised it. First flowers semi-double, later ones single. Sepals all the same length, which looks quite good. Colour is

'Perle d'Azur'

TOP: *C. × durandii* over *Senecio* 'Sunshine'
BOTTOM: *C. tangutica*

TOP: 'Lasurstern'
BOTTOM: 'Jackmanii Rubra'

TOP: *C. chrysocoma* var. *sericea* (syn. *C. spooneri*)
BOTTOM LEFT: 'Alba Luxurians'   BOTTOM RIGHT: 'Sir Trevor Lawrence'

TOP: *C. orientalis* in flowers and fruit
BOTTOM: 'Ernest Markham' and the Spanish broom *Spartium junceum*

TOP: *C. × eriostemon* with 'Huldine'
BOTTOM: 'Etoile Rose'

TOP: 'Etoile de Paris'
BOTTOM: 'Gravetye Beauty'

'Marcel Moser'

quite a dim magenta (C.L.), reasonable mauve (T.H.B.). Yellow stamens. Moderate vigour.

'Kathleen Wheeler' (B). Flowers terminal on longish young shoots. Leaves simple or, more usually, ternate with narrow, lanceolate leaflets. Flowers large, 16cm across; sepals 6, elliptical, rather gappy. A strong shade of rosy mauve or 'plummy mauve' but the 3 sunken veins in centre of each sepal are darker and pinker. Stamens creamy with a touch of lavender in the outer ones. A seedling from 'Mrs Spencer Castle', bred by Pennell's in 1952, after seed treatment with colchicine which may also explain its particularly large flowers.

'Keith Richardson' (B). 'Barbara Dibley' × 'Bracebridge Star' (1961) by Pennell's. Described by them as 'sharp petunia red with deeper bars, bold white centre'.

'Ken Donson' (B). Another Pennell cross: 'Barbara Jackson' × 'Daniel Deronda', made in 1961. 'Blue with golden anthers', say the raisers.

'Kermesina' (C). Vigorous cv of *C. viticella*. Of hybrid origin, and several forms going around. See *C. viticella,* p. 199. Also p. 81. Previous 'correct' name cv 'Rubra'.

'King Edward VII' (B or C). Leaves large, often simple and cordate; also ternate. Tends to carry solitary, terminal blooms on 60cm-long shoots of current season. Early blooms can be very green/grey and weird. Flowers 17cm across. Sepals 8, overlapping, obovate but with fine points; strong lilac-mauve with rosy flush in central area. Cream filaments, brownish anthers. A bold bloom but not prolific.

'King George V' (C). Vigour moderate. Leaves ternate. Flowers 15cm across. Sepals 6 or 7, flesh-pink with bright pink bar. Shy flowering.

C. koreana (C). Korea. Grows like *C. alpina,* to which it is related. A variable species. The name has erroneously been used as a synonym for *C. serratifolia.* The two species are not connected. Two varieties have been identified: var. *fragrans* has shoulders to the flowers; var. *lutea* has appendages, rather like nectaries. The latter is so like *C. serratifolia* that I believe it to be incorrectly named.

'Lady Betty Balfour' (C). Vigorous to 4m or more. Leaves usually ternate, shiny, coppery when young. Flowers 13cm across. Sepals 6, rather cupped, fine-pointed. Rich purple fading to bluer shade. White central bar on underside. Stamens creamy, conspicuous. September. Needs full sun.

'Lady Caroline Nevill' (B or C). Fairly vigorous. Leaves simple or ternate. Flowers 15–18cm across. Sepals 8, fairly pale lavender with very faintly darker midrib. Filaments white, anthers beige. Sometimes semi-double in first flush.

'Lady Londesborough' (B). Growth moderate. Leaves ternate. Flowers 13–15cm across, beautifully formed. Sepals 7, broad, round-tipped with wide overlap of margins. Pale mauve, fading to silvery mauve and finally palest grey. Prominent dark stamens.

'Lady Northcliffe' (B or C). Fairly vigorous once established. Slight coppery tinge in young leaves; simple, ternate or in fives. Flowers 15cm across. Sepals usually 6, broadly overlapping, wavy and never lying flat, much blunter than 'Lasurstern'. Deep, slightly richer blue than 'Lasurstern'. Stamens greenish-white, less prominent than those of 'Lasurstern'.

'Lasurstern' (B or C). Fairly vigorous to 3m. Leaves sometimes simple, usually ternate. Flowers 15–18cm wide, opening perfectly flat. Sepals 7 or 8 with wavy margins (cf. 'Lady Northcliffe', in which entire sepal is twisted), tapering to fine points. Rich, deep blue fading to campanula blue. Greenish-white bar on under surface. Stamens pale creamy. Second crop flowers only 13cm across.

'Lawsoniana' (B). Leaves simple or ternate. Flowers up to 24cm across, rosy lavender becoming lavender-blue. Tends to flower out of sight unless deliberately planted to be seen from above.

'Lilacina Floribunda' (B). Offered in a number of nurseries recently, it is, we believe, a synonym for 'Guiding Star' (q.v.).

'Lincoln Star' (B). Frequently little vigour but can make a good plant. Leaves ternate. Flowers 13cm across, shapely with 8 narrow, pointed sepals. Brilliant raspberry-pink almost to margins, though sometimes this colour is confined to a central bar and autumn blooms may be quite washy and uncharacteristic. Anthers crimson, showy.

'Little Nell' (C). Vigorous to 6m. Leaflets 5 or 7 and further subdivided.

Flowers 5cm across on 15cm pedicels. Sepals 4 or 6, recurved at margins. Broad outer band of pale mauve; white central bar. Stamens greenish.

'Lord Nevill' (B or C). Vigorous. Leaves simple or ternate, deeply bronzed when young. Flowers 15–18cm across. Sepals 8 (sometimes 6), broad and overlapping, very wavy-edged, tapering broadly to sharp points. Purplish-blue, especially purple towards centre-base. Veins conspicuously darker throughout the sepal and especially as the flower bleaches. Purple anthers form no contrast.

'Louise Rowe' (B). 'William Kennett' × 'Marie Boisselot'. Introduced by Fisk's in 1984. Raised by one of their customers, Mrs J. B. Rowe of Norfolk. When seen at Chelsea in 1985, the flower looked good, forming a rosette of pale mauve with all sepals a uniform size. Golden stamens. Not particularly large double flowers, but these are supplemented by singles and semi-doubles in its main crop in early summer.

C. macropetala (syn. *Atragene macropetala*) (A). Northern China, Siberia, 1910. Vigorous to 4m. Leaves biternate, serrate. Flowers nodding 8cm across, conspicuously and fully double. Sepals 4, 4cm long by 2cm wide, lavender with blue margins. Numerous slender staminodes, 2·5cm long; the outer 10–14 blue, the inner ones whitish. Cultivars include 'Blue Bird', 'Maidwell Hall', 'Markham's Pink', 'Rosy O'Grady' and 'Snowbird'. Others such as 'Anders', 'Harry Smith', 'Jan Lindmark', 'Pearl Rose' and 'Rodklokke' are offered mainly on the Continent and we haven't met them yet.

'Mme Baron Veillard' (C). Vigorous, to 4m in a season. Leaves ternate, each leaflet on a long stalk. Flowers rather cupped at centre but with slightly reflexed tips, 13cm across. Sepals 6, obovate, slightly gappy, a nice rosy-lilac shade. Stamens greenish-white.

'Mme Edouard André' (C). Not vigorous but a good doer, to 2·5m. Leaves simple or ternate. Buds narrow, pointed, flowers up to 13cm across, rather cupped. Sepals 6, very pointed; margins incurved (cf. 'Ville de Lyon'). Uniformly deep wine-red, matt surfaced. Stamens creamy.

'Mme Grangé' (C). Vigorous. Leaflets 5. Flowers 13cm across, freely borne. Sepals 4, 5 or (usually) 6; inrolled margins most pronounced on young plants. Upper surface deep purplish-red, velvety. Lower surface pubescent, dusky. Stamens grubby, nondescript.

'Mme Julia Correvon' (C). Vigorous. Leaflets 5, further subdivided. Flowers 13cm across. Sepals 4, 5 or 6, narrow and gappy with recurved tips, the upper surface deeply channelled. Clear rosy-red. Stamens greenish-cream. Very free.

'Mme Le Coultre'. See 'Marie Boisselot'.

'Maidwell Hall' (A). A bluer form of *C. macropetala*.

'Mammut' (B). From the USSR, raised by U. Kivistik. Created quite a stir when Jan Fopma showed a slide of it at a recent meeting of the International Clematis Society. White with a very pronounced boss of stamens. Raised in 1980, seed parent 'Mme Van Houtte', pollen parent unknown.

C. marata (A). Slender climber from South Island, NZ. According to Joe Cartman, the leaves are ternate, leaflets vary in size and shape. Flowers greenish, up to 6 per node, prominent bracts. 4 sepals often with twisted or rolled edge. Strong cinnamon scent.

'Marcel Moser' (B). Rather weak-growing, tending to flower at the expense of young growth. Leaves ternate, pubescent when young, longer and more narrowly tapering than those of 'Nelly Moser'. Flowers 20cm across. Sepals 8, tapering to fine points. Rosy-mauve with carmine central bar. Bleaches badly. Handsome reddish-purple stamens.

'Margaret Hunt' (C). See p. 52. Since my book's first edition, the eponymous Ms Hunt has sent me a spectacular photograph of this clematis in her garden, taken in July 1977. She cuts it back to about 50cm annually and it carries hundreds of sizeable pinkish blossoms on both sides of a screen running north and south.

'Margot Koster' (C). Robust and prolific. Leaflets 5. Flowers 8cm across. Sepals 4, 5 or 6, obovate, gappy with reflexed margins and tips. Rosy-red. Stamens whitish.

'Marie Boisselot' (syn. 'Mme Le Coultre') (B or C). Vigorous. Leaves simple or ternate, very broad. Flowers 15cm across, beautifully formed, opening flat. Sepals 8, very broad, overlapping so widely as sometimes to give semi-double impression. Faint pink flush at unfolding; soon pure white. Stamens white.

'**Markham's Pink**' (syn. *C. macropetala* 'Markhamii') (A). Similar to *C. macropetala* apart from its colour. Outer surface of sepals bright reddish-purple with narrow pale lilac margin. Inner, concealed staminodes, greenish-cream.

C. marmoraria (A). A New Zealand species, discovered December 1973. Low tufted sub-shrub up to 60cm across forming a tight little cushion. Spreads by runners underground. Leaves deep shiny green, leathery, hairy. Flowers solitary, white, flushed with green, male 2–3cm across, female 1·6–2·4cm. Smallest known clematis species. Chris Grey-Wilson, in *The Kew Magazine*, August 1987, remarked that it could be mistaken for a species of *Ranunculus* or *Anemone*. In the wild, found in only two areas of NZ, growing in crevices in limestone, above the treeline. First grown in UK as an alpine.

'**Maureen**' (B). Bushy, neither tall nor vigorous but free. Sepals 6, reddish purple (bluer than 'Mme Grangé'). A good colour and a nice clematis.

C. maximowicziana (syn. *C. paniculata*) (C). Japan 1860. Better known by its synonym, though *C. paniculata* is correctly the New Zealand species long known as *C. indivisa*. *C. maximowicziana* is a vigorous (more vigorous than the related *C. flammula*) climber to 10m. Leaves smooth, dark green on both sides with 3 or (more usually) 5 leaflets, edentulate, not even lobed and resembling, as Bean describes it, 'in form and colour the leaves of the common lilac'. Flowers cruciform, 3cm across, white, scented. Late season.

C. meyeniana (A). Hong Kong (*c.* 1821). Vigorous, evergreen armandii type. Pubescent stems and petioles. Leaves leathery, ternate, leaflets up to 10cm long. Flowers pinky-white or white, 2·5cm across, carried in clusters. Sepals usually 4, narrowly oblong. AM 1920. Prolific, tender, but good under glass perhaps.

'**Minuet**' (C). Vigorous. Leaflets up to 7, and further subdivided. Flowers 5cm across on long pedicels, each with two foliar bracts. Sepals 4, margined in deep mauve, the inner area off-white with mauve veins. Stamens green.

'Miriam Markham' (C). Moderate vigour. Large, loosely built, double flowers, shaded lavender. Shy flowering.

'Miss Bateman' (B). Fairly vigorous. Leaves ternate. Flowers 15cm across, opening flat and well formed. Sepals 8, broad, overlapping, creamy-white. Reddish-purple anthers make a striking eye.

'Miss Crawshay' (B). Vigour moderate. Leaves ternate. Buds very fat, opening into (usually) semi-double blooms, 15cm across. Single blooms have 8 sepals, round-tipped, narrow at base. Pale rosy-mauve. Anthers pale fawn.

C. montana (A). Himalaya. 1831. Exceptionally vigorous, to 12m. Leaves ternate with serrated margins. Flowers in axillary clusters up to 5, 5–8cm across, vanilla-scented. Sepals usually 4, white.

C. montana var. ***rubens*** (A). China. 1900. Leaves purplish, especially when young. Flowers rosy-mauve, 5–8cm across, vanilla-scented.

C. montana var. ***wilsonii*** (A). Central China. *c.* 1900. Flowers a month later than other montanas. Flowers 9cm across. Sepals 4, each 2cm across, white, rolling back along margins in older blooms. Prominent brush of stamens. Strong hot chocolate scent. Often sold untrue to name; Wilson himself introduced at least two clones.

The following cultivars of *C. montana* are some of those in commerce, but there are many more:

'Alexander'. Strong grower with large pale leaves. Creamy-white flowers up to 8cm across. Good scent.

'Elizabeth'. Large, light pinkish flowers, sometimes white if grown in shade. Scented.

'Freda'. Raised at Woodbridge in Suffolk by Mrs Freda Deacon, a seedling of *C. montana* 'Pink Perfection', recently introduced by Fisk's. Tom saw it in spring 1988, under glass at Harry Caddick's nursery in Cheshire. He confirms that it tallies with the photograph in Jim Fisk's catalogue and his accompanying description of 'deep cherry-pink with a dark edge to the sepals and bronzy foliage'.

'Grandiflora'. Flowers up to 8cm across with very wide white sepals. No scent.

'Marjorie'. A seedling of *C. montana* var. *wilsonii*, raised by Miss Marjorie Free of Westleton (Suffolk). Introduced 1980 by Fisk's. They describe it as 'semi-double, creamy-pink sepals with a centre of salmony-pink petaloid stamens'. The one I saw at Chelsea in 1984 was a muddy greenish, white-pink.

'Mayleen'. Introduced by Fisk's in 1984, given to him by a customer at Chelsea. 'Large deep pink with bronzy foliage.'

'Peveril'. Described by the nursery of the same name as 'a white with widely spaced sepals and long stamens'. It was recently collected in China and they say it flowers very late – mid to late July.

'Picton's Variety'. Less vigorous than average. Sepals 4, 5 or 6, of varying widths. Opens deep rosy-mauve (most intense in sunlight), fading only a little. Anthers whitish. Scent negligible.

'Pink Perfection'. Well scented, more rounded sepals and deeper colour than 'Elizabeth'. Several forms are going around under this name.

'Tetrarose'. Large rosy-mauve tetraploid cultivar. Nice cut leaf. Unfortunately several forms appear to exist.

'Vera'. Particularly vigorous. Large leaves with a pronounced purplish tinge. Well-scented, fair-sized flower. Good pink, though not so deep as 'Picton's Variety'.

'Moonlight' (B). Erroneously 'Yellow Queen' and 'Lawsoniana Henryi' – as explained in an article by Magnus Johnson published in the ICS *Journal* No. 1, 1984. A bit tricky and not too vigorous in cultivation, being liable more than most to wilt disease. Leaves ternate, each leaflet with a long petiole (especially the centre one). Leaf margins apt to be paler than central area. Flowers cream-white, retaining colour best in shade. Sepals 8, obovate, gappy. Stamens cream. A pretty thing.

'Mrs Bush' (B). Fairly vigorous. Leaves simple or ternate. Flowers 15–18cm across or more. Sepals 6–8, of soft, silky texture, fairly narrow. Uniformly pale bluish-mauve. Anthers pale brown.

'Mrs Cholmondeley' (B or C). Vigorous. Leaflets 3 or 5. Flowers 20cm across, prolific. Sepals usually 6, obovate, narrow at base and with gaps between sepals. Lavender-blue with darker reticulations in veins. Fades pleasantly. Stamens brownish. Long season. The blowsy old girl makes a splendid show.

'Mrs George Jackman' (B or C). Reasonably vigorous. Leaves simple or (more often) ternate. Flowers 15–18cm across with (usually) 8 broad overlapping sepals and sometimes an extra 2 or 3, giving slightly doubled effect. White. Anthers pale beige, forming a more conspicuous eye than 'Marie Boisselot' but less so than 'Henryi' and 'Miss Bateman'.

'Mrs Hope' (B or C). Vigorous. Leaves nearly all simple, up to 20cm long and 10cm broad in mature plants. Flowers 18cm across at first flowering, 13cm at second. Sepals 8, very broad (6cm across) and thus overlapping to give semi-double effect. Light blue with slightly darker central bar. Anthers rich purple. A different clematis is frequently sold as 'Mrs Hope'.

'Mrs James Mason' (B). From Peveril's. 'Vyvyan Pennell' × 'Dr Ruppel'. Described by the raisers as: 'double in its first flowering, single thereafter. Violet-blue with a red bar, cream stamens.'

'Mrs N. Thompson' (B). Like a very dark 'Barbara Jackman'. Sepals 6, overlapping. Petunia central bar with bluish-purple margins. Stamens reddish. Slow to get started.

'Mrs Oud' (B). Weak grower. Leaves ternate. Large white flowers, well shaped. Sepals broad and overlapping though pinched and upturned near base, leaving a gap here. Dark anthers.

'Mrs P. B. Truax' (B). Fairly vigorous. Leaves ternate. Flowers 14cm wide. Sepals usually 8, not much overlapping, very narrow and gappy at base. Mid to light blue, fading. Stamens creamy. Short season. Has been sold as 'Xerxes'.

'Mrs Robert Brydon' (C). Either a cv of *C. heracleifolia* or a clone of *C. × jouiniana* (*C. heracleifolia* var. *davidiana* × *C. vitalba*). Sub-shrubby to 2m, requiring good support. Leaves coarse, though not specially obtrusive. Flowers very pale off-white with faint bluish tinge, looking dirty. Sepals 4 or 5, entirely separate and gappy, rolled right back. Prominent white stamens.

'Mrs Spencer Castle' (B). Vigour moderate. Leaves ternate. Large, pale mauve-pink flowers with cream anthers. Sepals usually 6, but extra ones at first flowering, giving semi-double effect.

'Myojo' (B). From Japan via Fisk's. Purple, fairly large, 7–8 sepals and prominent white anthers. Not bad.

C. napaulensis (syn. *C. forrestii*) (A). Northern India, China. Introduced by Forrest, 1912. Vigorous evergreen, winter-flowering and somewhat tender species. Leaves ternate, untoothed or a few large teeth. Flowers in axillary clusters of up to 10, each subtended by a perfoliate bract (similar to that found in the related *C. cirrhosa*). Sepals 4, greenish-white, 1·5cm long, somewhat recurved at tips. Stamens 3cm long, protruding and conspicuous because purple throughout. No scent. Attractive seed heads. Sometimes incorrectly spelt *C. nepaulensis* or *C. nepalensis*.

'Nelly Moser' (B). Vigorous to 3·5m. Leaves ternate, smooth (cf. 'Marcel Moser'). Flowers 17cm across, opening flat. Sepals 8, fairly blunt-tipped, rosy-mauve with carmine bar. Brilliance of this bar varies greatly from one season to another. Bleaches badly. Anthers reddish-purple.

'Niobe' (C). For once an attractively and appropriately named new clematis. The one thing everyone remembers about 'Niobe' is that she was 'all tears'. Sepals 6. Lovely, very deep, velvety red. Lighter, greenish stamens.

C. ochotensis (A). China. Similar to *C. macropetala* but with shorter petaloid stamens. The Swedes have crossed it with *C. koreana* var. *fragrans* (as they have *C. macropetala* and *C. barbellata*), but we have not seen the results. *C. alpina* 'Frankie' is, we believe, a cross between *C. ochotensis* and *C. alpina*.

C. orientalis (syn. *C. graveolens*) (C). Asia. 1731. Vigorous to 5m. Size and bright yellow colouring of flowers similar to *C. tangutica*, but the 4 sepals often opening wide to 5cm, though sometimes campanulate. Found by Ludlow and Sherriff in 1947, L & S No. 13342 in a good form (it was introduced from seed and tended to be variable from the first) used to outclass all previous discoveries of this widely distributed and variable species. Leaflets up to 9 and further finely subdivided. Flowers broadly campanulate with thick, fleshy sepals. Stamens conspicuous, with reddish-purple filaments and white anthers. It has been outclassed by *C. orientalis* 'Bill Mackenzie', which has larger flowers and more vigour. Unfortunately, seedlings appear to have been distributed; we have seen at least four forms purporting to be the original. 'Corry', 'Gravetye Seedling' and *sherriffii* are all listed as forms of *C. orientalis* which, together with *C. tangutica*, is itself now considered to be a sub-group of *C. vernayi*, which now appears to be a sub-group of *C. tibetana*. May the Saints preserve us!

'Pagoda' (C). One of John Treasure's. *C. viticella* × 'Etoile Rose'. Aptly named: narrow sepals, strongly recurved. Pinky-mauve with faint bar. Referred to as a texensis type but owes more to *C. viticella*. Strong grower. A bonny plant.

'Pamela Jackman' (A). A selection of *C. alpina*. Flowers 8cm across at mouth. Sepals 4·5cm long, 2cm wide; of generous proportions for this species. Rich, deep azure. Staminodes in a tight cluster, the outer ring bluish, the inner ones creamy.

C. paniculata (syn. *C. indivisa*, by which name it is still more familiar in this country) (A). New Zealand, abundant in both islands and their showiest species. 1840. Evergreen, dioecious, the males being showier than females. Leaves ternate, the leaflets entire or lobed as is said to be the case in the cv 'Lobata', though my specimen had no lobes. Flowers 8cm across, borne in axillary panicles. Sepals 6–8, spathulate, pure white. Stamens yellow.

C. parviflora (A). Vigorous evergreen New Zealander with a wide distribution along lowland forest margins as far south as Nelson. Small, delicately divided leaves, usually with 5 pinnae, which are themselves subdivided. Flowers in axillary clusters of 2, 3 or 4 making dense ropes of blossom. Each flower 4cm across. Sepals 6, lanceolate, making a star shape. Colour throughout, including stamens, a slightly yellow shade of green. Pretty but not eye-catching. Can be strongly scented of lemon. Variable species in wild. According to Joe Cartman (Canterbury Botanical Society *Journal* No. 19, 1985), it should now be called *C. cunninghamii*.

'Pennell's Purity' (B). Raised in 1962 by Pennell's, 'Beauty of Worcester' × 'Marie Boisselot'. Often semi-double, described as 'warm white with crimped sepals and golden stamens'. Not widely available.

'Percy Lake' (B). Large flowers. Sepals 7 or 8, washed-out mauve-white; stamens the same.

'Percy Picton' (B). Fairly weak-growing. Leaves ternate. Flowers 18cm across. Sepals usually 8 (sometimes only 6 or 7), perfectly ovate, a rich intense mauve. Anthers purple.

'Perle d'Azur' (C). Vigorous to 4m. Leaflets 3 or 5. Flowers up to 15cm across, but, with recurved tips, appearing less. Sepals 6 (occasionally fewer), broadly obovate, 7cm across at broadest. Deeply channelled midribs. Light blue with pinkish-mauve flush towards centre-base. Stamens pale greenish. Very free.

C. petriei (A). From New Zealand, where it is found over small trees and shrubs or scrambling over rocks. Ternate leaves, bright green above, paler below. Prolific. Flowers prolific, green, 4·5cm across in male. Strong scent. Striking seed heads on females. Fairly hardy in UK. Now considered a sub-species of *C. forsteri*.

'Peveril Pearl' (B). Offered only by Peveril's, as far as we know. They describe it as 'a large flower, lilac with a pink-flushed midrib, cream stamens with violet tips. Vigorous and free-flowering.'

C. phlebantha (A) is as fully described as I intend, p. 60. Also written up in the RHS *Journal*, 1968.

'Phoenix' (B). From Magnus Johnson, raised around 1958. He describes it as 'of rather rough growth up to 2m. Free-flowering. Large flowers 15 to 23cm across, violet with a petunia purple bar, white anthers tipped purple.' Not currently available in UK.

'Pink Fantasy' (B or C). Obtained from Canada by Fisk's, who introduced it in 1975. Much too pale (pink), and although described as semi-double the extra sepal spoils the flower. Dark anthers. Compact grower to 2m.

C. pitcheri (syn. *C. simsii*) (C). USA. 1878. Climber to 3m. Leaflets 3, 5 or 7 and further subdivided. Flowers solitary, urn-shaped. Sepals 4, violet outside, purple within, 3cm-long with recurved tips. Related to *C. texensis*, with which it has been confused by French hybridizers.

C. potaninii (C). Very similar to *C. fargesii* var. *souliei* but with larger flowers.

'President, The'. See 'The President'.

'Prince Charles' (C). Alister Keay recalls that this was given to him back in the 50s when his New Zealand clematis business was in its infancy. Introduced into UK by Fisk's in 1986. No pedigree available. Flowers June to September. Mauvish (blue blood transfusion desperately awaited). Described as a compact grower, 1·5 to 2m. Any aspect except north.

'Princess of Wales' (B). An old Jackman cross (according to Spingarn). Returned to UK and Jim Fisk's list in 1986 after a long sojourn in New Zealand, where it was previously grown by Alister Keay. Mauve, we believe. Named after Queen Alexandra.

'Prins Hendrik' (syn. 'Prince Henry') (B). Weak growing but enormous flowered. Leaves usually simple. Sepals 7, broad, with crimped, wavy margins and fine points; lavender-blue. Anthers purple. Best under glass.

'Proteus' (B). Moderate vigour. Leaves trifoliate. Buds globular. Flowers fully double in first crop; pompon rosettes 15cm across. Sepals about 100, rosy-lilac. Second crop flowers, single; sepals 6, darkest at margins. Stamens cream.

C. quadribracteolata (A). As neither of us had seen this, we had decided to leave it out. A pity; we rather liked the name. Serendipity intervened. Later that day we saw it in Jack Elliott's greenhouse (Kent). The first of the New Zealand species to flower, it has tiny greenish-brown leaves. The flowers, we gather, are small, brown or purple. Sepals narrow, 15mm by 2mm.

C. quinquefoliata (A). Central and west China. Late-flowering member of armandii group, closely related to *C. meyeniana*. Leaves pinnate, leaflets 5–10cm long. Downy, ribbed stems. 4–5m. Flowers white, 4–5cm across in clusters, 4–6 sepals. August–September. Said to have striking seed heads.

'Ramona' (syn. 'Hybrida Sieboldiana') (B). A large, lavender-blue clematis, very close to 'Blue Gem', 'Mrs Bush' and 'Mrs Hope'. 'Lady Caroline Nevill' has also been seen as this!

C. recta (syn. *C. erecta*). Europe and Asia. 1597. Herbaceous climber. Height variable; 1 to 2m. Deep green pinnate foliage; leaflets 5. Flowers in panicles, white, 2cm wide, cruciform, sometimes scented, mid-summer-flowering. Conspicuous seed heads. The cv 'Purpurea', with purple young foliage maturing green, is not a fixed clone.

C. rehderiana (syn. *C. nutans; C. nutans thyrsoidea*) (C). Western China. 1898. Vigorous to 6m. Leaves and shoots very downy; leaflets 7 or 9, coarsely toothed. Flowers in axillary panicles, bell-shaped, 1·5–2cm long. Sepals 4, recurved at tips, straw yellow. Strongly cowslip-scented.

'Richard Pennell' (B or C). A clematis of Walter Pennell's raising, Richard being his son. The seed parent is 'Vyvyan Pennell' (in human terms also), pollinated by 'Daniel Deronda' (ahem). Sepals 8, with strongly waved margins. Lavender with whitish intimations. Very prominent whirling stamens with red filaments which curve over at the tips; cream anthers. A pretty flower at all stages and an outstanding clematis.

'Rosy O'Grady' (A). *C. macropetala* × *C. alpina*. Raised 1964 by Dr F. L. Skinner at Dropmore in Canada. Vigorous. Light rose sepals, longer than the type. Repeat blooms through the summer. 3m.

'Rosy Pagoda' (A). A selection of *C. alpina*. Pale pink, a good doer.

'Rouge Cardinal' (B). Introduced by Jim Fisk from the Continent. An excellent clematis. Magenta flowers up to 15cm across, blunt-edged and of a velvety texture. Easy. Grows to 3m.

'Royal Velours' (C). Vigour moderate, considering its viticella blood. Leaflets 5 and further subdivided. Flowers 8cm across. Sepals usually 4, dark rich purple, of velvety texture.

'Royalty' (B). Raised by John Treasure. 'Maureen' × 'Countess of Lovelace'. Double similar to 'Vyvyan Pennell' but a darker blue-purple. Single flowers in autumn. Quite compact, moderate vigour.

Rubro-marginata. See *C.* × *triternata* 'Rubro-marginata'.

'Ruby' (A). A selection of *C. alpina*. Flowers up to 8cm across. Sepals 4cm long, of a dim lilac colouring. Staminodes off-white with mauve tinge.

'Saturn' (B). A Pennell hybrid, 'Lord Nevill' × 'Nelly Moser'. Sown 1963, but named in 1978 after the death of Walter Pennell whose favourite hobby was astronomy. An attractive name, it is described as 'lavender with a maroon bar, white stamens with dark tips'.

'Scartho Gem' (B). From Pennell's. 1963. 'Lincoln Star' (seed parent) pollinated by 'Mrs N. Thompson'. Bright, rosy carmine with narrow, paler sepal margins. Beige stamens. Some semi-doubles in first flush. Eye-catching. Named after one of the Pennell nurseries.

'Sealand Gem' (B or C). Fairly vigorous. Leaves simple or trifoliate. Flowers 13–15cm across. Sepals 6, perfectly ovate, 4cm across. Pale rosy-mauve with carmine bar quickly fading to give uniform colouring. Anthers mauvish, not standing out. Main flowering on young shoots.

'Serenata' (B or C). Raised in Sweden by Tag Lundell in 1960, a seedling of 'Mme Edouard André'. Flowers 10cm across, larger (up to 15cm) from old wood, plum-purple with a deeper bar and yellow stamens. Quite vigorous to 3m. Young growth conspicuously hairy.

C. serratifolia (C). Korea. *c.* 1918. Vigorous woody climber to 4m. Leaves biternate, smooth though surface often puckered. Coarse, uneven teeth. Flowers usually in threes, the centre one opening earliest, nodding open-bell-shaped with somewhat spreading segments, 4cm across at mouth. Sepals 4, up to 1·5cm wide, gappy, pale yellow with tinge of green. Stamens very distinctive being purple throughout. Notable lemon scent.

'Silver Moon' (B or C). Rather unkindly noticed on pp. 37–8, but vigorous and bushy to 3m and has a long season. Excellent on a north or other sunless exposure which is, indeed, the only position where its colouring is acceptable.

'Sir Garnet Wolseley' (B). Moderate vigour. Leaves nearly all ternate (occasionally entire); leaflets up to 7cm across, 9cm long, tapering to fine points. Flowers 13cm across. Sepals 7 or 8, obovate, tapering at base to leave gaps. Colour good at all ages, velvety purple with reddish flush at first, fading to blue-mauve. Anthers reddish purple.

'Sir Trevor Lawrence' (C). Vigorous semi-herbaceous climber. Leaflets 3, 5 or 7. Flowers shaped like upright bells, 5cm long. Sepals up to 6, very much the same cherry-red as 'Gravetye Beauty' but slightly more luminous. Fades to bluer shade. Stamens cream throughout (cf. 'Grave-tye Beauty's' red anthers).

'Snowbird' (A). From Peveril's, a seedling of *C. macropetala* 'Markham's Pink'. Pale foliage. White flowers, rather like *C. alpina sibirica* 'White Moth' but larger, they recurve attractively at tips. Flowers later than other macropetalas.

'Snow Queen' (B). From Alister Keay (New Zealand). A good white with a hint of mauve at margins. 6 sepals, pointed; deep reddish anthers. Sepals crimped, rippling at margins. Good. The best clematis shown at Chelsea 1986.

C. songarica (C). Non-climbing sub-shrub to 2m. Leaves simple, toothed, up to 11cm long, 3·5cm across, very smooth, shiny and green. Stems ribbed and grooved. Inflorescence terminal, consisting of about 30 blooms. Flowers 3·5cm across but tips of sepals rolled back. Sepals 5 (or 4), pure white, hawthorn- or meadowsweet-scented. Introduced 1880 from S. Siberia, Mongolia, etc., and the region of the river Sungari, from which it takes its name.

C. spooneri. See *C. chrysocoma* var. *sericea*.

C. stans Herbaceous. Height 1m, non-clinging, flopping. Leaves up to 18cm long, ternate. Central leaflet 3-lobed, about same size as lateral leaflets. Flowers monoecious, 2cm long, tube slender, not swollen at base, opening into 4 segments which curl back at tips. Blue-tinged white.

'Star of India' (C). Vigour moderate. Leaflets 3 or 5, broad and bold. Flowers 14–16cm diameter, of excellent shape with 4, 5 or 6 broadly ovate sepals, up to 9cm wide. Purple with broad reddish-purple central bar which gradually fades out. Stamens grubby, nondescript. Handsome and dashing.

'Susan Allsop' (B). 'Beauty of Worcester' × 'King Edward VII'. Pennell's, who raised it in 1962, describe it as: 'Rosy purple with magenta bar and golden stamens.'

'Sylvia Denny' (B). Raised by the Dennys of Preston and named after Mrs Denny. 'Duchess of Edinburgh' × 'Marie Boisselot'. Semi-double clear white, without any trace of greening. Nicely shaped rosette. Single flowers on young wood. A compact grower to 2·5m.

C. tangutica (C). China. 1898. Vigorous to 5m. Leaflets up to 7, unevenly toothed. Flowers campanulate, 2·5–4cm long. Sepals 4, butter-yellow inside and out. Conspicuous seed heads with feathery styles. The variety *obtusiuscula*, introduced by Wilson from W. Szechwan, China, in 1908; also a selection from it, 'Gravetye Variety', have smaller, deeper yellow flowers. 'Aureolin' (from Holland) is a good form, recently introduced. 'Warsaw' is another with which we are not familiar.

The President (B or C). Vigorous. Leaves simple or ternate; young foliage bronzed. Flowers up to 18cm across, somewhat cupped and readily showing under surface. Sepals 8, overlapping with tapered points. Upper surface uniform light purple; lower surface pale with silvery-white central bar. Anthers reddish-purple.

C. tibetana subsp. **vernayi** (C). The new grouping for those formerly called *C. orientalis*, including L & S and the many going around as 'Orange Peel'. *C. tibetana* can be seen offered as *thibetanicus* and *thibetianus*; no doubt other variations exist. The clone we have seen had finely-cut glaucous foliage, with purple stems. Flowers: thick, pubescent, lime-green sepals, dark anthers. Wide opening. Strong grower. Maybe not fully hardy. We call a halt here; the subspecies list goes on for ever.

'Titania' (B). A seedling of 'Nelly Moser' raised by Magnus Johnson. He gave a description in an article for the International Clematis Society *Journal* in 1984: Leaves ternate. Flowers 12–18cm across. 8 sepals, white with faint violet shading towards the base, maroon anthers. A moderate grower to 2·5m.

C. × triternata 'Rubro-marginata' (syn. *C. × rubro-marginata, C. flammula* var. *rubro-marginata*) (C). A cross between *C. flammula* and *C. viticella*. Vigorous to 4m. Flowers 2·5cm across. Sepals 4–6, white centred, shading to purple margins. Scented like *C. flammula*.

'Twilight' (C), but flowering rather early on its young wood like 'Comtesse de Bouchaud'. Plenty of vigour. Leaves simple or ternate. Flowers 11cm across, almost carmine at first with greeny-yellow stamens in marked contrast. Colour soon fades pleasantly to rosy mauve, with stronger rosy colouring maintained in central bar for some time. Stamens yellow. Sepals 6, broad, blunt (no points) and overlapping. A good flower, inappropriately named.

C. uncinata (syn. *C. leiocarpa*) (c). China. Evergreen with pinnate or biternate foliage. Flowers in axillary panicles from current season's growth. About 4cm across with 4 narrow, white sepals. Slight but not agreeable scent. Requires wall protection. Easily raised from seed. Introduced by Wilson, 1901. AM 1922.

C. × *vedrariensis* (*C. chrysocoma* × *C. montana rubens*) (A). Vigorous once established, to 6m. Leaves trifoliate, downy, deeply serrated. Flowers rosy-mauve, 7–8cm across on 10cm, downy stalks. Sepals 4, 5 or 6. Cv 'Rosea' (syn. *C. spooneri rosea*) is very similar. 'Highdown' is a selection of *C.* × *vedrariensis*.

C. veitchiana (c). Very close to *C. rehderiana* but the leaves are bipinnate, not pinnate. Leaflets 20 or more. More numerous but smaller than in *C. rehderiana*. Bracts on the inflorescence small and awl-shaped (larger and ovate or oval in *C. rehderiana*). Introduced by Wilson from W. China, 1904.

'Venosa Violacea' (c). Vigorous. Leaflets 5, subdivided into threes or fives. Viticella-type, but with 10cm-wide, upward-facing flowers. Sepals 5 or 6 with purple incurved margins and white, purple-veined centres. Stamens abortive.

'Venus Victrix' (B or c). Although not at present on offer, as far as I know, there are still a few plants around claiming to be this old cv for which Cripps, the raisers, obtained an FCC in 1876. The specimen I saw was under glass at Carolside, Berwickshire. Growth coarse. Leaves usually simple, sometimes ternate, cordate. Flowers terminal, not freely borne, large, semi-double with 2 rows of sepals, approximately 8 sepals in each row. Clear lavender-mauve. Filaments white, anthers purplish. Moore and Jackman give this as a double clematis with 6 or 8 rows of sepals. This could well be so on an outdoor specimen. Their description otherwise tallies with the notes I made on the spot. Not unlike 'Miriam Markham'.

'Veronica's Choice' (B). A Pennell clematis named after his daughter. The seed parent is 'Vyvyan Pennell', pollinated by 'Percy Lake'. Fully double, the centre of the flower pale mauve or even French grey; the outer, guard sepals a little more intense. I like this.

C. verticillaris (syn. *Atragene americana*, *C. occidentalis* var. *occidentalis*) (A). North America: Hudson Bay to Utah but now rare. 1797. Woody climber to 3m. Leaves ternate. Flowers solitary, nodding, 5–8cm wide. Sepals 4, 4cm long, purplish-blue as also outer staminodes. This is the North American counterpart of *C. macropetala* and *C. alpina* from the old world. Var. *columbiana* (syn. *C. columbiana*) is commoner in cultivation but not easy.

'Victoria' (C). Vigorous. Leaflets 5. Flowers 15cm across, opening flat, without recurving. Sepals 6 (sometimes 4 or 5), with deep central veins and minutely puckered surface, broadly ovate, 5cm across. Light rosy-purple with more intensely flushed central bar. Fades pleasantly. Buff stamens.

'Ville de Lyon' (B or C). Vigorous to 4m. Leaflets 3 or 5. Flowers in two big crops. Maximum diameter 14cm; 8–10cm later. Sepals 6, obovate, broad and blunt-tipped with margins and apex curved back. Deep carmine on opening, fading in centre to mauve while margins remain dark. Stamens creamy.

'Violet Charm' (B or C). Offered by Jim Fisk, who had it from a nursery near Birmingham. Beyond that, origin unknown. Described as 'violet blue, long pointed sepals with crimped edges and beige stamens'. Grows to 2·5m.

'Violet Elizabeth' (B). 1962. 'Vyvyan Pennell' × 'Mrs Spencer Castle'. Pennell's, the raisers, describe it as: 'Fully double mauve pink flowers formed like a double "Miss Crawshay".'

C. viorna (C). Eastern USA. 1730. Herbaceous climber to 2m. Leaflets 3, 5 or 7, their tips acting as tendrils. Flowers solitary, almost globular but for recurved sepal tips which give overall urn-shape. Sepals 4, very thick and succulent, 2cm long, dusky reddish. Seed heads green and conspicuous.

C. vitalba, Traveller's Joy, Old Man's Beard (C). Europe, including Britain; North Africa, Caucasus. Rampant climber to 10m. Leaflets 5, coarsely toothed. Flowers 3cm across in panicles. Stamens more prominent than the 4 sepals, all being dingy creamy-white with greenish undertones. Little or no scent. Seed heads woolly and conspicuous, persisting in winter.

C. viticella (c). Southern Europe. Vigorous climber to 3 or 4m. Leaflets 7, further subdivided. Nodding purple flowers 4–9cm across. Sepals 4, often recurved at margins. Pedicels 15cm long, thread-fine (with foliar bracts), giving graceful appearance to the shrub in bloom. Cv 'Purpurea Plena Elegans' (syn. 'Elegans Plena') is a strong-growing hybrid with flowers 5cm across, packed with sepals in a dense rosette. Soft 'old rose' purple. The vigorous cv 'Kermesina' is also of hybrid origin. Leaflets 7. Flowers outward-facing. Sepals 4 (or 5), 3cm wide, deep velvety crimson, white at base. Styles very dark red, almost black. Anthers brownish. Formerly misnamed 'Rubra'.

'Voluceau' (c). Reddish-purple flowers. 5 or 6 sepals and pale stamens. When I saw it at Sissinghurst, I noted that it was not very exciting and didn't show up well.

'Vyvyan Pennell' (b). The most vigorous large-flowered double clematis. Leaves ternate. Flowers form neat rosettes, 15cm across. Frame of large lilac guard sepals. Cushion of smaller lavender-blue inner sepals. A few single flowers on young wood. Walter Pennell made this cross between 'Daniel Deronda' and 'Beauty of Worcester'.

'Wada's Primrose' (b). Similar to 'Moonlight' and similarly prone to wilt. Does not have the long leaflet stalks of the above. Flowers even paler. Possibly another clone of the same patens sub-species or hybrid.

'Walter Pennell' (b). Mr Pennell described this as a companion to 'Vyvyan Pennell', which is the seed parent pollinated by 'Daniel Deronda'. The overall colouring is deep lilac, redder than 'Vyvyan' in the outer row of guard sepals which have a deep carmine bar. Centre of flower greyish, showing pale stamens when fully open but looking better before this stage is reached. One does not name a flower after oneself without thinking highly of it, but my own reaction is that we have here a clematis inviting comparison with 'Vyvyan Pennell' yet without striking a new chord. Time will tell.

'Warsaw Nike' (c). From Brother Franczak via Fisk's. Named after a memorial in the Polish capital to freedom fighters. Fisk's describe it as a 'royal velvet purple with golden stamens, free flowering and vigorous to 4m'.

'W. E. Gladstone' (b or c). Not vigorous, though with very long internodes and capable of reaching 3m. Leaves simple or ternate. Leaflets large, up to 9cm across and 15cm long. Upper surface very shiny when young. Flowers up to 25cm across. Sepals 6 or 7 (though Moore and Jackman give 8), obovate, uniformly lavender. Stamens prominent; anthers purple.

'White Columbine' (A). A white selection of *C. alpina*. We have not yet clapped eyes on it. There are a number of unnamed white seedlings of *C. alpina* going around, many of which look quite good.

'White Moth' (A). A selection of *C. alpina* var. *sibirica*. Light, yellowish-green, biternate foliage. Flowers 6cm across, white throughout with hint of green. Sepals 3·5cm long, 1·5cm wide. First row of 4 staminodes 3cm long, projecting beyond sepals but very narrow; inner staminodes conspicuous and numerous. Hence a fat, fully double flower resembling *C. macropetala* but smaller. Latest-flowering of its group.

'Wilhelmina Tull' (B). Pedigree unknown. Produced by a customer of Jim Fisk. He describes it as 'an improved "Mrs N. Thompson", deep violet with a crimson bar and golden stamens. Grows to 2·25m.'

'Will Goodwin' (B). Pale lavender, with 6–10 broad, 7cm-wide sepals, overlapping, corrugated, wavy margined. Cream stamens. Not unlike a smaller version of 'Prins Hendrik'.

'William Kennett' (B or C). Vigorous. Leaves nearly all simple. Flowers 15–18cm across. Sepals 8, broad and overlapping, rough-textured with crimped margins. Strong lavender-blue with reddish-purple central stripe on opening (far more marked than in 'Mrs Hope'), gradually fading out. Dark purple anthers.

'Willy' (A). A selection of *C. alpina*. A free-flowering plant. Pale pink sepals with a red blotch at the base of each. 2·5m.

'Xerxes'. The true plant is no longer in cultivation. Of recent years 'Elsa Späth' and 'Mrs P. B. Truax' have both been erroneously listed as 'Xerxes' in this country, while in Australia a white cv has for years been sold under this magnetic name.

'Yellow Queen'. See 'Moonlight'.

Chapter 13
NAMES: THEIR SPELLING
AND PRONUNCIATION

'Ascotiensis' was received by Wyevale Nurseries without a name. A plant without a name is no use to a nurseryman so they called it 'Wyevale Blue'. Only subsequently did they discover . . . but by then it was too late to go back, as any nurseryman will appreciate. (The general public may not.)

'Bees Jubilee'. There is no apostrophe anywhere in Bees. If there ever was one it has long since atrophied. I am always forgetting this and may have slipped one in here and there in the text.

'Bouchaud'. 'Comtesse de Bouchaud' is correct but you as often see Bouchard. I asked Hugh Thompson which was correct and he originally inclined to Bouchard, since this was an old family and 'royal or aristocratic patronage played such a part in naming Clematis in the past. My earliest reference agrees with yours, a reference to Bouchaud in some notes Morel wrote for an early edition of Robinson's famous book. I observe though that Markham who clearly knew Morel and many other French growers used Bouchard. The first Comte de Bouchard lived in the tenth century, and was variously known as "le Vénerable", "le vieux" and "Chauve Souris". His daughter burned herself to death in public apparently, without making it quite clear why she took this somewhat drastic action. Later the Bouchards secured the Vendôme title before being appointed Comtes Royal de Paris. In 1371 the 7th Comte married Isabella de Bourbon, and from then appeared to be absorbed by the Bourbons. I have not checked whether the Bouchard title has been used by minor Bourbons, or the offspring of those that had committed indiscretions, the usual fate of secondary titles in continental Europe. It would be interesting today to flush a de Bouchard from the inner recesses of the French Jockey Club.'

That was on 28 March 1969, but a little later that spring he wrote in a p.c. 'Evidence for BOUCHAUD almost clinched in my opinion. I was re-examining the evidence for Morel's claim that "Ville de Lyon" was a texensis hybrid (you will recollect that he claimed it was produced by crossing coccinea with "Viviand Morel") and I found he sent "Ville de Lyon" to Monsieur le Comte de Bouchaud "l'amateur passioné d'horticulture" who grew it and flowered it in his garden at Chasselay (Rhône) for several years before it was made available to commerce. Both Edouard André and de Bouchaud added their support for Morel's claims to its ancestry, for seemingly inconclusive reasons. This unexpected light on the subject makes me suspect that when he produced this pink clematis he named it after de Bouchaud's wife between 1900 and 1906. I am still baffled at the complete lack of recognition of any Bouchaud Title in books that specialize in such matters.' This reference to M. le comte de Bouchaud is in *Revue Horticole*, 1899, p. 274.

'Campanile'. Pronounced Campeneely.

'Capitaine Thuilleaux' is often seen in British growers' lists as Capitan, which means a swaggerer or bully. I thought this an odd epithet, and suggested in my last book that he had been demoted when he crossed the Channel. A few years ago Hugh Thompson wrote to John Treasure that Thuilleaux was a French nurseryman who specialized in raising variants of 'Nelly Moser' in between the wars. He apparently named this one in memory of his father, who was killed in the First World War. Unless Thuilleaux had an exceedingly bad father, Capitan must surely be Capitaine. Some growers compound the error by calling it 'Souvenir de Capitan Thuilleaux'.

'Cholmondeley'. Pronounced Chumley. I remember overhearing the musings of a woman visiting my garden when she contemplated a defunct but still labelled specimen. 'Mrs Chol-mon-deley. Mrs Chol-mon-deley is dead. Poor Mrs Chol-mon-deley.'

Clematis. The correct pronunciation is, unequivocally, clĕm'atis, with a short e and the accent on the first syllable. All the dictionaries are agreed on this, and even Fowler's *Modern English Usage*, in a (to me) impenetrable article on False Quantity, comes down in favour.

But 'This climber's common fate is
   To be pronounced clemā'tis',

with the accent on the second syllable and a long ā. In America clemă'ttis is common, rhyming with lattice.

The Continentals have a hard time of it. Magnus Johnson (a Swede) confided to Tom that he had to be very careful while in the UK, lest he commit the heresy of saying 'clemartis' (the long a is the norm over there). What an intolerant lot we are – the poor man sounded quite intimidated.

Most nurserymen in this country use clĕm'atis correctly: Pennell, Fisk, Picton, Treasure, Jackman, to wit, though the Saunders family of Knight's Nurseries prefer clemā'tis. So did Margery Fish, while Percy Thrower, as another sinner, must have had a considerable influence on box viewers. When he brought his programme to my garden he kindly tied in with my pronunciation. In return, I referred with him to the grey foliage plant as *Cineraria maritima* instead of *Senecio cineraria* as I normally should have. What's known as being flexible.

For the plural of clematis I (and many others) use the same word. I cannot say this is correct, but if enough of us go on doing so for long enough it will become correct. The English language is overburdened with sibilants, especially in the final syllable or syllables. Crocuses and irises are tolerable, but clematises leaves the tongue congealed and torpid. Imagine 'she possesses 66 clematises'. Most unpleasant.

The word clematis has both Latin and Greek origins, though the *OED* tells us that it referred to some climbing or trailing plant unspecified but probably periwinkle.

'Correvon'. G. S. Thomas used to refer to the plant at Hidcote Manor, of which he had a good colour transparency, as 'Mme Jules Correvon'. Through his great influence and the trust we all place in him, this gained currency, following the cultivar's emergence from long eclipse. Unfortunately, when some of us had delved into early records, it transpired that the much less easily spoken 'Mme Julia Correvon' is correct.

'Elsa Späth'. We can all pronounce Elsa but Späth is the devil if you've no Teutonic foundations. It's something like Shpāte, with a long ā. If English printers cannot rise to the German umlaut (the ¨ over the a) which makes it a long vowel, they should write it Spaeth (as they do for the lilac, 'Souvenir de Louis Spaeth'), never Spath, which is inevitably pronounced with a short ă.

'Etoile Violette' will never sound the same again. Some time ago, Tom overheard someone tell a colleague that he had this clematis and it was doing well, whereupon the other person looked puzzled and asked what had a clematis to do with property leases? Was the name a joke? Following a lengthy conversation at cross-purposes (in a noisy public bar) the matter was resolved. The listener had been convinced the plant was called 'A Twelve Year Let'.

'Eva Maria'. I know nothing about this clematis but cannot help wondering if it didn't start as 'Ave Maria' and undergo a sea change.

'Fair Rosamond' not Rosamund.

'Gipsy Queen' not Gypsy.

'Gravetye Beauty'. Pronounced Grave Tie not Gravity, it is the name of William Robinson's home near East Grinstead in Sussex.

'Huldine'. A French name, roughly Hooldeen.

'Jackmanii'. When English names are latinized it is best for the English to pronounce the surname as near as possible to the way you would in ordinary speech and then tack on the necessary suffix. Thus, having discovered that *Halesia* is named in honour of a person of the name of Stephen Hales, you can never again say Haleesia. It's different with foreign names like Fuchs giving rise to *Fuchsia*. Foreigners (including Scots like Menzies) must look after themselves. Hell! I've done my best for them in this book, haven't I? There are limits.

In this case the house of Jackman is being celebrated. Just say Jackman and then tack on a long i. So simple. If you must hiccough, I allow you Jackmannyi. But what we so often hear is Jackmarny. 'I'm so tired of it,' I told Mr Jackman, once. 'So am I,' he agreed, emphatically.

'King Edward VII' is the name, not Edward VII (Treasure's please note) or it'll be Teddy VII or King Tum-Tum before we know where we are. King Edward the Seventh.

'Lasurstern'. A German name. My Dutch gardener pronounces it with ease and grace – so naturally. I do envy him; I'm always being asked how it should be said and invariably succumb to hoarseness and spluttering. Stern is German for a star and Lasur presumably means azure. Azure Star. Not a bad name. How about it?

Macropetala. Say the word so that it means what it is trying to say. Macro = big. Petal = petal. Then *say* petal and not Macropetarla.

'Madame Baron Veillard'. On nurserymen's labels the title is often shortened to Mad. This can cause consternation; Tom once overheard a lady, in all seriousness, say to her husband: 'Oh, what a terrible thing to have done! Fancy calling a plant after a mad Baron.'

'Nelly Moser' not Nellie. It was always Nelly from the first report in the *Revue Horticole* in 1897.

'Nevill' as in 'Lord Nevill', 'Lady Caroline Nevill', the family that lives at Eridge near Tunbridge Wells. They doubtless patronized Cripps (later Wallace of the Old Gardens) at Tunbridge Wells and were honoured by having new clematis named after them. There is no e terminating the name as in Lord Rupert Neville.

'Niobe'. Pronounce the final e as you would in Hebe, Phoebe, Nerine and Irene (let's hope).

'Lady Northcliffe' *not* Northcliff (the sort of cliff that faces north).

'Mrs Oud' is pronounced Owd. A Dutch word meaning old and a common surname. We seem to prefer Young in this usage.

'Pennell', Vyvyan, Walter, Richard and descendants to the last trump. Pronounce with the accent on the first syllable as in Fennel (the herb).

'Perle d'Azur', not Perle D'Azure. If we're going to stick to the French (and we do) then it is no use semi-anglicizing. The only possible alternative is to go the whole hog with Azure Pearl, as also

'Prins Hendrik', which can be translated into Prince Henry and very welcome, too. But don't try to be authentic and get it wrong, e.g. Prins Hendrick.

'Royal Velours' not Velour. Velour derives from the old French velous, which in turn is supposed to derive from Villosus (hairy). The English word, of course, is velvet.

Sepal. Pronounce with a short ĕ as you do petal. Not seeple.

'The President' is correct, though often shortened to 'President' *tout court*. I have been guilty of this myself, as you never know, where the definite article is included, whether to look the word up, alphabetically, under 'President, The' or 'The President'. However, as Mr Rowland Jackman rightly observed, 'The President' was the name that Charles Noble gave to the plant and 'The President' it is, like it or not.

'Mrs P. B. Truax'. A dreadful name, admittedly, but not improved by being misspelt Traux.

'Ville de Lyon'. Lyon has to be spelt the French way, without a terminal s, otherwise the only alternative would be City of Lyons.

'Warsaw Nike' is named after a memorial in the Polish capital and is pronounced Neekay.

'William Kennett' has and always had two t's. Not to be confused with Lord Kennet, whose name having made news has probably caused the confusion.

'Wolseley'. Sir Garnet Wolseley was the gentleman's name; no relation to Cardinal Wolsey.

'Xerxes'. Don't try and pronounce the first letter as a separate syllable: Exerxes. Pronounce it as a Z, Zerxes. In any case there is no such clematis, but Pennell's have applied this good selling name to 'Elsa Späth' and Fisk has fallen from grace and followed suit.

Glossary
of Botanical Terms Used

ANTHER. The part of the stamen containing the pollen grains.

AXIL. The junction of leaf and stem. AXILLARY. Arising therefrom.

BIPINNATE. Of a leaf in which the primary divisions are themselves pinnate.

BITERNATE. Of a leaf divided into three parts which are themselves divided into three.

BRACT. A modified leaf growing near the flower.

CARPEL. One of the units of the female part of the flower.

CORDATE. Heart-shaped.

CULTIVAR. A plant which has originated in cultivation. Abbreviated to cv.

DIOECIOUS. Having the sexes on different plants.

FILAMENT. The stalk of the anther, the two together forming the stamen.

FOLIAR. Leaf-like.

GENUS. A group of species with common structural characters. The name of the genus is, in designating a plant, placed first and has a capital initial letter.

GLABROUS. Without hairs.

GLAUCOUS. Bluish.

HERMAPHRODITE. Containing both stamens (male) and ovary (female) in the same flower.

LANCEOLATE. Narrowly elliptical, tapering at both ends.

MONOECIOUS. Having unisexual flowers, but both sexes on the same plant.

NODE. A point on the stem where one or more leaves arise. Adjective NODAL.

OBOVATE. Broadest above the middle (of a leaf).

OVATE. Broadest below the middle (of a leaf).

PANICLE. An axial flower stem along which the flowers are arranged on branching stalks.

PEDICEL. The stalk of a single flower.

PEDUNCLE. The stalk common to a branch of flowers.

PERFOLIATE. Having the stem apparently growing through the leaf where 2 opposite leaves are united at the base.

PERIANTH. The floral leaves as a whole, including petals and sepals if both are present.

PETALOID STAMENS. Stamens which are petal-like in colour and texture.

PETIOLE. The stalk of a leaf.

PHOTOSYNTHESIS. The synthesis, made by the green plant parts, of sugar and starch from carbon dioxide and water, in the presence of daylight.

PINNATE. Of a leaf composed of more than three leaflets arranged in two rows along a common stalk.

PUBESCENT. Shortly and softly hairy.

SEPAL. One of the parts forming the outermost whorl of the flower.

SERRATE. Toothed like a saw.

SPATHULATE. Paddle-shaped.

SPECIES. A group of individuals which have the same constant and distinctive characters. The name of the species is, in designating a plant, placed second, and has a small initial letter.

STAMEN. One of the male reproductive organs of the plant.

STAMINODE. An infertile stamen.

SYNONYM (SYN.). A superseded or alternative name.

TERNATE. Of a compound leaf divided into three parts.

VARIETY (VAR.). A natural group within a species, occurring in the wild; normally given a name of Latin form.

WHORL. More than two organs of the same kind arising at the same level.

× The sign for hybridization.

Index

For the benefit of the general reader, where there are many page numbers under one heading, **bold** type is used to indicate more than just a passing reference to the subject.

US Mail-order Sources for Clematis

CARROLL GARDENS
PO Box 310C, Westminster MD 21157.
(Many clematis. Catalog $2.00 deductible.)

CLIFFORD'S PERENNIALS AND VINES
Route 2 Box 320, East Troy WI 53120.
(Small-flowered clematis.)

THE COMPLEAT GARDEN
177 Argilla Road, Ipswich MA 01938.
(Limited mail-order. Inquire with SASE.)

FORESTFARM
990 Tetherow Road, Williams OR 97544.
(Several species clematis.)

D. S. GEORGE NURSERIES
2515 Penfield Road, Fairport NY 14450.
(Large selection of large-flowered hybrid clematis.)

LOUISIANA NURSERY
Route 7 Box 43, Opelousas LA 70570.
(Southern plant varieties including
Clematis virginiana, *C. crispa* and *C. texensis*.)

THE SEED SOURCE
Route 2 Box 265B, Asheville NC 28805
(Seeds for species and large-flowered clematis.)

WAYSIDE GARDENS
PO Box 1, Hodges SC 29695–0001

WHITE FLOWER FARM
Litchfield CT 06759–0050